W9-CRZ-203

New Membranes and Advanced Materials for Wastewater Treatment

ACS SYMPOSIUM SERIES **1022**

New Membranes and Advanced Materials for Wastewater Treatment

Anja Mueller, Editor
Central Michigan University

Benoit Guieysse, Editor
Massey University

Abhijit Sarkar, Editor
Michigan Molecular Institute

Sponsored by the
ACS Division of Environmental Chemistry

American Chemical Society, Washington DC

Library of Congress Cataloging-in-Publication Data

New membranes and advanced materials for wastewater treatment / Anja Mueller, editor.
    p. cm. -- (ACS symposium series ; 1022)
    "Sponsored by the ACS Division of Environmental Chemistry."
    Based on presentations given during the spring 2008 ACS Meeting.
    Includes bibliographical references and index.
    ISBN 978-0-8412-7214-9 (alk. paper)
    1. Sewage--Purification--Reverse osmosis process. 2. Water--Purification--Materials. 3. Membranes (Technology)--Materials. I. Mueller, Anja. II. American Chemical Society. Division of Environmental Chemistry. III. American Chemical Society. Meeting (235th : 2008 : New Orleans, La.)
    TD754.N49 2009
    628.3'52--dc22

                                2009039506

Foreword

The ACS Symposium Series was first published in 1974 to provide a mechanism for publishing symposia quickly in book form. The purpose of the series is to publish timely, comprehensive books developed from the ACS sponsored symposia based on current scientific research. Occasionally, books are developed from symposia sponsored by other organizations when the topic is of keen interest to the chemistry audience.

Before agreeing to publish a book, the proposed table of contents is reviewed for appropriate and comprehensive coverage and for interest to the audience. Some papers may be excluded to better focus the book; others may be added to provide comprehensiveness. When appropriate, overview or introductory chapters are added. Drafts of chapters are peer-reviewed prior to final acceptance or rejection, and manuscripts are prepared in camera-ready format.

As a rule, only original research papers and original review papers are included in the volumes. Verbatim reproductions of previous published papers are not accepted.

ACS Books Department

Contents

Part II Advanced Membranes

Indexes

Preface

It is a pleasure to write this introduction for "New Membranes and Advanced Materials for Wastewater Treatment." Professor Anja Mueller and her colleagues, Dr. Benoit Guieysse and Dr. Abhijit Sarkar, have edited a stimulating collection of papers based on presentations given during the Spring 2008 ACS Meeting.

Mid-summer 2008, BCC Research released a market research report for membrane bioreactors (MBRSs). Their research suggests the global MBR market will grow at a compound annual growth rate of 10.5% from 2008 to 2013. Based on this prediction, sales in 2013 would reach $488 million of which ~40% would be for systems in municipal and domestic wastewater treatment.

The MBR industry is roughly two decades old having grown from $1 million in 1990 to $298 in 2008. The phenomenal growth of this industry is due to the energy efficiency and process intensification (higher flow rates, smaller footprints) offered by MBR technology. Moreover, MBR systems produce a high quality effluent from wastewater that can be safely discharged to the environment or reused for industrial or human consumption.

BCC predicts slightly slower growth for membranes in all liquid separations - 7.7% from $2.1 billion in 2008 to $3 billion in 2013. This includes separations ranging from particle filtration to reverse osmosis. The growth in the industry has led to announcements of significant production line expansions by most manufacturers, especially in Asia.

Growth of the industry is due in part to manufacturers incrementally improving production processes over time. Additionally, initial problems with membrane lifetime and fouling were solved to an acceptable level.

The impact of these improvements is dramatically evident in desalination. Desalination by reverse osmosis is an order of magnitude more energy efficient than by thermal processes, such as multi-stage flash, and water production costs are 25% less - a win-win situation!

Continued growth will require new, innovative solutions to fouling which creates the largest headaches for membrane users. Additionally, as membrane processes produce higher water purity, concerns over lower concentration contaminants (e.g., hormones, drugs, metals, toxins, and other small molecule contaminants) must be addressed.

This volume reports state-of-the-art research into these issues. Several chapters describe the synthesis and use of molecularly imprinted polymers - exciting new materials that can selectivity remove compounds at very low concentrations. A number of chapters describe innovative materials and processes for specific metal removal and concentration.

Additionally, promising new approaches to fouling reduction though fine control of pore structure, biorepellent coatings, and high throughput screening of surface modifiers are described.

I believe the membrane community will find the work described here thought provoking and research stimulating. I highly recommend it to you.

Professor G. Glenn Lipscomb
University of Toledo, USA

Chapter 1

Introduction: Recent Progresses in Materials Science for Water Reclamation

Anja Mueller[1]*, Benoit Guieysse[2], Abhijit Sarkar[3]

[1] Department of Chemistry, Central Michigan University, Mt. Pleasant, MI 48859, E-mail: muell1a@cmich.edu
[2] School of Engineering and Advanced Technology, Massey University, Private Bag 11 222, 4442 Palmerston North, New Zealand
[3] Michigan Molecular Institute, 1910 West St. Andrews Road, Midland, MI 48640-2696

In 2007, the Public Utility Board (PUB) of the Republic of Singapore won the prestigious Stockholm Industry Water Award for its holistic approach to water resources management. Interestingly, PUB champions the use of advanced water treatment technologies such as membrane filtration and has long preferred the term "used water" to "wastewater", a concept that better reflects the nature of its business and vision. This story illustrates a global trend in the water treatment sector that is further evidenced by the US$ 2.5 billion venture capital reportedly invested in US Clean Technologies in 2007, a 45% increase from 2006 (PricewaterhouseCoopers).

Increasing pressures to reduce operating costs, improve treatment performance, and minimizing environmental impacts are thus progressively reshaping the water treatment industry into a product recovery business. This generates a need for novel resource efficient technologies that must still remove target toxicants from increasingly complex cocktails of chemicals whilst optimizing the qualitative and quantitative output of its products in the form of reclaimed water, nutrients, energy, or even high value chemicals.

With this perspective, this volume demonstrates that recent innovations in material sciences provide tremendous potential to develop a novel generation of water treatment technologies.

1

Chapter 2 and 3 thus describe the use of selective molecularly imprinted adsorbents for the removal of trace contaminants such as endocrine disruptors, a technology potentially enabling high-value chemical recovery without wasting removal capacity on harmless substances present at high concentrations. These adsorbents still function as coagulation particles or filter media as well.

Toxic metal ions cannot compete in binding to the coagulation resins, in this case because other ions are several times more abundant in water. Again, these resins can be made more specific by imprinting polymerization, significantly increasing the capacity of these resins. Chapter 4 describes an example that was developed for both hydrophilic and hydrophobic metal complexes.

Chapter 5 describes a different method to create specific binding sites for toxic metal ions, specifically lead and hexavalent chromium. In this case, common monomers are polymerized in the presence of chelating ions, forming strong, specific binding sites. In Chapter 6, on the other hand, water purification uses an inexpensive and abundant clay as a filtering material to remove heavy metal ions. Copper removal was optimized by characterizing the zeolite crystal structure and choosing the most effective mixture of structures.

The next chapter goes a step further: the material chosen does not only remove toxic metal ions, but it also detects them. This is especially useful in cases when toxic metals are introduced into water only intermittently and without warning.

Finally, Chapter 8 describes an example for the potential of materials science for product recovery; Struvite, a fertilizer, is isolated from dairy wastewater, reducing the amount of nutrients in the remaining water at the same time.

The second part of this book covers the materials science of membranes. Membrane processes cover a wide range of applications from waste water treatment to medical applications. For some applications, membranes play an important role in which other technologies are not capable such as medical sector (hemodialysis) and energy sector (fuel cell). Once considered a viable technology only for desalination, membrane processes are increasingly employed for removal of bacteria and other microorganisms, particulate materials, and natural organic material, which can impart color, odors, and tastes to the water and react with disinfectants to form disinfection byproducts. As advancements are made in membrane production and module design, capital and operating costs continue to decline. A vast array of products is being treated using membranes, but water desalination is using over 80% of all membranes having ever been sold. Membrane processes have become more attractive for potable water production in recent years due to the increased stringency of drinking water regulations. Membrane processes have excellent separation capabilities and show promise for meeting many existing and anticipated drinking water standards. While all types of membranes work well under proper conditions, choosing the most appropriate membrane for a given application still remain crucial.

The demand for pure process water, clean potable water and acceptable wastewater treatment continues to grow. Microfiltration (MF), ultrafiltration (UF), nanofiltration (NF), reverse osmosis (RO) and desalination (D) processes are widely used for many municipal and industrial applications and replace some existing conventional systems that cannot deliver pure water. Membrane Separation is also used in regions where freshwater is a precious commodity and where many industrial plants are recovering their own water.

During the past two decades, RO process has gained extensive attention in reclamation of water and separation of organics from aqueous streams. However, successful utilization of RO technology has been hampered by fouling, which poses a major obstacle for the membrane application. Therefore, there is a growing interest in surface modification of existing RO membranes to introduce properties that markedly reduce fouling, specifically biofouling, while retaining water flux and rejection characteristics relative to conventional RO membranes. Commercial RO membranes are typically made of thin-film-composite aromatic polyamides for which the microorganism-induced biofilm formation on the surface is a major problem. This biofouling increases the operating pressure by about 50% thus requiring regular cleaning by chlorine treatment. In fact, apart from the cost of energy to run the high-pressure pumps, membrane fouling is the single most important factor that controls the cost of RO water purification unit (ROWPU) operations. It leads to an undesirable reduction of the flux of permeate water by the source water contaminants during the course of operation and therefore, requires cleaning them every 4 to 8 weeks and changing the membrane elements approximately every 3 years. Since the maintenance and remediation expenses represent ca. 30% of the total operating cost, a new generation of membranes with inherent anti-fouling capability is urgently needed. The world is still waiting for a good composite membrane for RO and NF which can tolerate e.g. 20 ppm sodium hypochlorite.

The possibility of developing new nanostructured materials with specific configurations and morphology is offering powerful tools for the preparation of membranes with controlled selectivity and permeability higher than the membranes existing today. Membranes characterized by highly selective transport mechanisms as the perovskite studied for oxygen separation from air, or the palladium for H_2 purification are suggesting the use of molecular dynamic studies for identifying new structures characterized by similar selectivity towards a larger spectrum of chemical species. Biological membranes reproduce themselves continuously, controlling important physiological processes, where fouling e.g. does not represent a problem as in artificial systems. The mechanisms which generate our memory or the function of our brain are other important membrane phenomena. The role that membrane science and membrane engineering play in our life, justifies growing efforts in the education of young generations of researchers, engineers and technicians on their basic properties and on their possible applications.

Materials science and engineering can help in the solution of the difficulties, such as specificity, ease of fabrication of controlled-structure membranes, and biofouling. The following chapters are several examples of solutions for these problems.

Chapter 9, for example, presents theory and examples of how to fabricate membranes with an absolute control over pore size, shape, and three-dimensional arrangement. With these membranes it is now possible not only to separate particles, but also to fractionate them into different size fractions.

These controlled membranes, though, still suffer from biofouling, as all membranes do. The next chapter, therefore, describes how these membrane surfaces can be modified to reduce fouling. Grafting from the surface is used for the surface modification; easily implementable syntheses were developed for a wide variety of surface coatings. Thus this is a flexible method that can be used for a wide variety of membrane and water remediation applications.

Another class of polymeric materials, hyperbranched and dendritic polymers, can be used as anti-fouling coatings as well. With this class of materials it is possible to also structure the coating on the nanoscale, as well as encapsulate compounds that can give the coating additional functionality, such as bactericidal properties. Chapter 11 and 12 explore different examples of such a coating.

It is difficult and very time-consuming to design, synthesize, and test anti-fouling coating materials. Chapter 13, though, describes a method that will speed up the development of anti-fouling coatings considerably: it describes a high-throughput synthesis and screen for anti-fouling coating materials. With this method it will be possible to quickly identify several promising candidates, which could then be developed into functional coatings for a variety of applications.

The last chapter shows an example where filtration is combined with photocatalytic oxidation to regenerate membranes continuously. The membranes include titanium(IV)oxide nanowires that destroy small particles and bacteria on the surface of the membrane in the presence of UV or sun light.

The materials described in this volume are only a few examples of the potential of materials science applications for water reclamation. With further development, these innovations will lead the path to the future of the Water Industry and more globally, the Clean Technologies Sector.

Part I:
Advanced Materials

Chapter 2

Removal of Endocrine Disrupting Compounds Using Molecularly Imprinted Polymers: A Review

Benoit Guieysse[1,2], Mathieu Le Noir[3,4], and Bo Mattiasson[3]

[1]School of Civil and Environmental Engineering, Nanyang Technological University, Block N1, Nanyang Avenue, Singapore 639798 Singapore
[2]School of Engineering and Advanced Technology, Massey University, Private Bag 11 222, 4442 Palmerston North, New Zealand
[3]Department of Biotechnology., Lund University, PO Box 124, S-221 00 Lund, Sweden
[4]Anoxkaldnes AB, Klosterängsvägen 11A, SE-226 47 Lund, Sweden

Various independent studies have recently demonstrated the potential of Molecular Imprinted Polymer (MIPs) as selective adsorbents for the highly efficient removal of EDCs at trace concentration. Current water purification techniques are often inefficient at trace concentration because substances interfering in the removal process are present at higher concentration. Instead, molecular imprinting is based on the same mechanism that makes EDCs so harmful: their capacity to bind to natural receptors, which also makes it possible to remove unknown compounds with endocrine disrupting activity. Molecular recognition is the basis for natural molecules (proteins) to specifically bind to certain substances at extremely low concentrations. In comparison to classical adsorbents, the efficiency of imprinted polymers can be maintained in the presence of interfering matter, and regeneration can be easily achieved by solvent extraction under mild conditions. MIPs are also more stable than bio-adsorbents (antibodies, receptor proteins) and can be reused in many cycles. They can finally be synthesized in a vast variety of shape and size. Further research is needed to improve the group selectivity of imprinted materials.

"Stay calm everyone, there is Prozac in the drinking water!" (1) - Playing upon the satirical side of the alarming presence of anti-depressors in drinking water, this recent headline in the British Press illustrates one of today's largest environmental threat: the increasing environmental occurrence of trace contaminants that can be toxic at trace concentration due to the repeated exposure of target populations and the cumulated effects from the many substances simultaneously found (2-5).

Endocrine Disrupting Contaminants (EDCs)

Among the trace contaminants found in the environment, endocrine disruptors are certainly among the most worrisome (6). EDCs include substances of wide ranges of properties, sizes and molecular structures that are able to disrupt the endocrine system of animals in various ways, such as mimicking natural hormones (inducing wrongly timed signals or binding to and blocking natural hormone receptors - repressing the natural hormone action). Unfortunately, a surprising variety of man-made chemicals are capable of endocrine disruption and the large scale ecological effects of endocrine disruptors represent an enormous risk. The old toxicity dogma which states "the dose makes the poison" is no longer true as these compounds show potency at extremely low levels (Table 1).

Table 1: Concentration of endocrine disruptors in releases from wastewater treatment plants

Compound	Concentration (ng/l), location
17β-oestradiol (E2)	2.7-48, UK (7); 0.44-3.3, Italy (8); 3.2-55, Japan (9); 4.5-8.6, France (10) 1.1, Sweden (11); 6, Canada (12); 0.2-4.1, USA (13); 0.3-2.5, Japan (14); 3-8, Italy (15); 1.6-7.4, UK (16)
17α-ethynyloestradiol	2.7-4.5, France (10); 9 , Canada (12); 1, Germany (12); 4.5, Sweden (11); 3, Germany (17); 0.6, Italy (8); 1.4, Italy (18)
Estrone	3, Canada (12); 9, Germany (12); 2.5-34, Japan (14); 5-30, Italy (15); 6.2.-7.2, France (10); 5.8, Sweden (11); 6.4-29, UK (16)
Nonylphenol	200, Japan (19); 6600, Spain (20); 2900, Canada (21); 300, Japan (22); 730, Belgium (23); 370-530, Italy (24); 100-1000, Japan (24)
Bisphenol A	380, Spain (25); 1000-2700, Canada (26); 20-100, Japan (24); 4000, the Netherlands (27); 1400, USA (28)
Diclofenac	900, Spain (25); 40, South Korea (29); 2510, Germany (30); 145, Japan (31); 90, USA (32)
Endosulfan	12100, Germany (33);
Lead	51000, Germany (33); 500, India (34); 2700, Greece (35); 31100, Italy (36)

It is precisely their capacity to cause harmful effects at very low concentrations that make many EDCs such a problem. There is sometimes a confusion between 1) endocrine disruptors, which share a toxicity mechanism, 2) trace contaminants, which represent compounds found at concentration below 1 mg/l (typically 1 µg/l – 1 ng/l), 3) emerging contaminants, which are compounds that were only recently found in the environment and 4) pharmaceuticals and personal care products (PPCPs) which regroup contaminants based on their use. In fact, many PPCPs are endocrine disruptors that were only recently detected in the environment at trace concentration. PPCPs are especially problematic however because many drugs are purposely designed to be biologically active at very low concentrations and are often not regulated as hazardous substances. Pharmaceuticals administered to patients will be excreted as such or in metabolized forms primarily via the urine and will thus end up in low concentrations in wastewater. These pharmaceuticals are not sufficiently degraded during wastewater treatment and are therefore able to enter the environment. They are also released to the environment in a dispersed and diffuse form that is much harder to control. Thus, although most EDCs are intrinsically not persistent in the environment, their constant release can offset their natural attenuation, leading to their pseudo-persistence and potential chronic biological effects on humans and wildlife.

Technical limitations to current wastewater and water treatment methods

The first extensive study on the occurrence of drugs in sewage treatment plants and rivers was presented in 1998 (37) and based on their results, the authors concluded that in respect of their ubiquity, many of these substances should be considered as environmental pollutants. Soon after, the EU and the US EPA initiated major programs to map the environmental occurrence of PPCPs and reached similar conclusion (38-40). The US Geological Survey for instance found that endocrine disruptors were frequently detected in US river streams at worrisome total concentrations although the individual concentration of species was often low (28). Similar compounds were found in European surface waters (41).

Table 2: Type of treatment and removal efficiency for some EDCs from wastewater influent

Country	Compound	Process type	Removal efficiency	Ref.
Brazil	Oestrone	Trickling filter	67%	(12)
Brazil	Oestrone	Activated sludge	83%	(12)
Italy	Oestrone	Activated sludge	74%	(18)
Italy	Oestrone	Activated sludge	61%	(8)
Italy	17β-oestradiol	Activated sludge	87%	(8)
England	PCB	Activated sludge	96%	(42)
Switzerland	Nonylphenol	High loading/nitrifying	37%	(43)
Switzerland	Nonylphenol	Low loading/non-nitrifying	77%	(43)
USA	17β-oestradiol	Sand filtration	70%	(13)
England	Triazines	Conventional 2 stage	<40%	(44)

As most emerging trace contaminants are released in the environment from wastewater treatment discharges, the ubiquity of these compounds in surface water bares the testimony that classical wastewater treatments are inefficient for their removal (Table 2). Numerous studies have thus shown that:

- Classical processes (coagulation, flotation, and biological treatment) can only partially remove trace pollutants such as EDCs (45, 46).

- Advanced processes (activated carbon adsorption, ozonation, advanced oxidation, membrane filtration) are efficient for many pollutants but not all of them. They are also considerably limited by the competitive removal of interfering organic matter which wastes most of the adsorption or oxidation capacity. As a typical example, it was necessary to introduce 5.0 mg L^{-1} of O_3 to remove 2 μg L^{-1} of nonylphenol in river water, which is almost 800 times the amount theoretically needed to fully mineralize this pollutant (47). Similarly, Fukuhara et al. (48) observed that the adsorption capacity of activated carbon for E2 dropped 1000 times to 0.14-0.20 μg g^{-1} from 0.53-4.1 mg g^{-1} when 1 μg E2/L was supplied in river water instead of water due to site competition and/or pore blockage (the river water contained approx. 11 mg/L of total organic carbon).

Economical and societal limitations to EDCs control

It is likely that a combination of source control (by for instance improving the collection of old medicines, reducing use abuses and improving dosage) and use of several advanced technologies for wastewater and drinking water treatment would bring strong improvements. However, this will come at a huge cost and will unlikely completely prevent the release of endocrine disruptors in the environment for various reasons:

1. Phasing out many of the compounds involved is impossible. DDT is still largely used to avoid malaria nearly 50 years after its environmental effects were discovered, saving thousands of lives annually. The same dilemma will have to be faced for many of the drugs we use.
2. Pollution control is virtually impossible when the sources are diffuse, which is the case for many emerging pollutants.
3. Many "reservoirs" of pollutants (i.e. contaminated soils, sediments, or waste dumps) will continue to supply EDCs to surface waters for many years.
4. Certain compounds need to be endocrine disruptors (i.e. birth control agents).
5. The in-depth assessment of all potential harmful chemicals is prohibitive and unrealistic, given the fact that many compounds are unknown. There are today 500 known biochemical receptors at which drugs are targeted and this number is expected to increase up to 20-fold in the near future (2).
6. The environmental impact of the use of advanced treatment to remove trace contaminants is questionable in regards to the increase in energy consumption and CO_2 emmision this would generate (49).

In conclusion, it is currently very difficult to reduce the risks of EDCs exposure because their release cannot be prevented and current treatment methods are inefficient to remove them at an acceptable cost. Conventional methods are intrinsically limited by their lack of specificity at trace concentration because most of the adsorption, biodegradation, or oxidation capacity is wasted (Table 3). Consequently, the efficiencies of current methods are highly unpredictable for the hundreds of new chemicals commercialized every year and the even greater number of unknown pollutants (i.e. metabolites). New treatment methods capable to bring EDCs concentrations down to extremely low levels are therefore crucially needed.

Synthesis and characterization of molecular imprinting polymers

Molecularly Imprinted Polymers (MIPs) are synthesized by guided polymerization of functional monomers around a target molecule (the template), which leaves a specific recognition site after removal of the template (Figure 1): By using an endocrine disruptor as template, the adsorbing material becomes a synthetic analogue to the natural receptors that can remove any molecule having the capacity to bind natural receptors (i.e. any potential endocrine disruptor). Thus, removal is ingeniously based on the same property that makes the pollutants so harmful: their capacity to bind to specific natural receptors.

Molecular recognition is the basis for natural molecules (proteins) to specifically bind to certain substances and thereby, for the living organism to detect the target compounds at extremely low concentrations. Synthetic receptor analogues obtained via molecular imprinting can be used in many of the applications earlier developed based on natural receptor molecules such as the construction of biosensors (67), packing material for bioseparations (68) and for the extraction and purification of biological and environmental samples (69). Molecular recognition is highly specific and highly sensitive and can allow the enrichment of trace contaminants without the problem of competitive extraction of interfering substances.

Molecular Imprinting

Molecular imprinting is a technique used to induce molecular recognition properties in synthetic polymers in response to the presence of a target molecule acting as a template during the formation of the three-dimensional structure of the polymer (70). Two basic approaches can be distinguished based on the interactions used in the imprinting step:

- The covalent approach, in which functional monomers and templates are bound to each other by covalent linkage prior to polymerization. This covalent conjugate is then polymerized and the imprinted molecule is removed by chemical cleavage (70).
- The non-covalent approach, in which the pre-polymerization arrangement of the template and the functional monomers is formed by non-covalent interactions such as ionic interactions or hydrogen bonding (70). Following polymerization, the template can be removed by solvent extraction. The

12

principal means of re-binding the target molecule to these polymers is achieved again through non-covalent interactions (70).

The non-covalent method is quite simple, easy to perform and control, and is by far the most widespread in the literature (71). Both approaches have advantages and disadvantages but in general, noncovalent imprinting shows a greater choice of functional monomers, target templates and imprinting materials (70, 71). The most suitable method for a given application depends on the needs and conditions (template, selectivity required, cost and time available for preparation, etc.). The conceptually and practically simple "self-assembly" non-covalent approach is similar to natural processes by which hormones operate (72, 73). Hence, the non-covalent approach has so far been preferred for the imprinting of hormonally active organic compounds.

MIPs synthesis typically requires a template (i.e. the target contaminant); a monomer, a cross-linker, a solvent (also called porogen) and an initiator. The type and amount of each compound must be optimized in view of the desired selectivity, stability, reproducibility, and cost-efficiency. 17β-oestradiol (E2) is often chosen as model EDC for being considered as the most active estrogen (74) and for being frequently detected in wastewater (75).

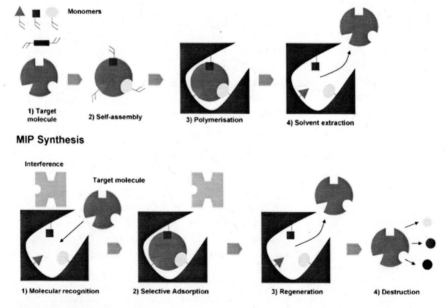

Figure 1: A target molecule, for instance an endocrine disrupter, is introduced during the polymerization process of appropriate monomers. These polymerize to form a polymer cage. The print molecule can then be dissociated from the polymer imprint. In the presence of the MIP, the target pollutant is selectively adsorbed without poisoning by interfering molecules. The MIP is then regenerated by solvent elution of the pollutant that is subsequently destroyed.

Table 3: Advantages and limitation of conventional technologies for trace contaminant removal

Technology	Advantages and limitations
Biological processes	Economical and environmentally sound, biological processes are limited at low concentration as 1) micro-organisms prefer to metabolize other organic compounds present at higher concentrations, 2) trace contaminants are not detected or not present at concentrations high enough to be useful as substrate (50, 51). More generally, the mechanisms of substrate uptake and microbial growth at trace substrate concentrations are still poorly understood (52). Several studies have demonstrated the potential or extra-cellular fungal oxidases (laccase) for the removal of EDCs at trace concentrations (53-60). However some limitations include that 1) extra-cellular oxidades are usually not very substrate-specific; 2) economical large-scale production of fungal oxidase remains difficult.
Adsorption	Although efficient for hydrophobic contaminants, adsorption is very limited for soluble of semi-soluble contaminants and when interfering substances are present (i.e. humic acid, surfactants) (48, 61). Regeneration of activated carbon is also difficult and costly (48, 62).
Ozonation	Ozonation is efficient to remove certain contaminants but surprisingly fails for others (63, 64). Its main disadvantages are 1) high capital and operating costs, 2) lack of specificity (large amount of O_3 are wasted for the oxidation of interfering material), and 3) risks associated with the release of hazardous intermediates.
Advanced Oxidation Processes	Although very efficient for a large range of contaminants, these methods are expensive and limited by the presence of interfering substances which must be oxidized before free radicals can attack the target contaminants.
Membrane filtration	The efficiency of microfiltration and ultrafiltration is limited to hydrophobic pollutants adsorbed to particles whereas nanofiltration and reverse osmosis are efficient for more compounds (65, 66). Unfortunately nanofiltration and reverse osmosis are not specific (they remove most water constituents) and carry high capital and operation costs.

Template and functional monomer

More than 200 compounds have been tested as templates in non-covalent molecular imprinting protocols. Compounds that are not chemically inert under polymerization conditions are not suitable for non-covalent imprinting (76) and imprinting of biological macromolecules, such as proteins, is still under development. The functional monomers are responsible for the binding interactions at the imprinted binding sites, and the choice of monomer is dependent on the required template characteristics (77). Acidic, neutral and basic monomers can all be employed in non-covalent imprinting, but by far the most widely used monomers include methacrylic acid and vinylpyridine.

The ratio of template to functional monomer is a crucial parameter affecting the performance of the imprinted polymer. A template:functional monomer ratio of 1:4 is commonly applied during non-covalent imprinting protocols and provides a complex of sufficient stability (77). Too little functional monomer may result in few imprinted sites, while too much may result in too many non-specific binding sites, although significant imprinting effects have been noted with a template:monomer ratio of 1:500 (78). A large amount of wasted template may cause template leaching issues (77, 79) but can be overcome by synthesizing a MIP based on a structural analogue of the target moelcule, i.e. a dummy template (80).

Crosslinkers

The fundamental roles of the crosslinker are to 1) stabilize the imprinted binding sites in the polymer 2) control the morphology and stabilize the polymer matrix, and 3) regulate the porosity and the hydrophobicity of the polymer to facilitate its practical application (60). Ethylene glycol dimethacrylate (EDMA) is a commonly used crosslinker (61, 81). A sufficient amount of the crosslinker is required to be effective. EDMA has mostly been used at approximately 80% of the total molar amount of monomer + crosslinker in the polymer (81). If the concentration of the crosslinker is too low during polymerization, it will not be possible to fix the template molecule to form three-dimensional molecular imprinting. On the other hand, too much crosslinker may hinder the re-binding of the target molecule to the binding sites (82).

Solvent and initiator

Three major characteristics regulate the choice of the solvent used in the polymerization process. First, the solvent should have the ability to solubilize the precursors before polymerization (72, 83). Second the solvent, or "porogen", should promote the formation of pores in the polymer matrix (77) in order to facilitate the release of the bound target molecule from the polymer (70) and later provide a large surface area for pollutant adsorption. Increasing the volume of the porogen increases the pore volume (76). The solvent also influences the pore diameter, and therefore the surface area (70). Solvents employed in noncovalent imprinting protocols are generally non-polar/aprotic, such as chloroform, acetonitrile or toluene. They facilitate non-covalent interactions by stabilizing hydrogen bonds (84). The most commonly used initiators are 2,2'-azobis-(2,4-dimethylvaleronitrile) and 2,2'-azobisisobutyronitrile (AIBN).

Polymerization procedure

Imprinted polymers can be obtained in various shapes or sizes depending on the polymerization protocol used (71, 79). In the most traditional approach, polymers monoliths are formed that must then be powdered and sieved. Alternatively, discrete polymer microspheres can be achieved using precipitation polymerization, which basically follows the same protocol with the difference that a much larger amount of sovent is used (85, 86).

Polymerization is generally performed with reaction times between 16 and 48 hours, depending on the batch size and the format. Polymerization is typically induced by UV light irradiation or heat (45-120°C) (87). The choice of the polymerization process depends on the template reactivity towards UV or heat stability. Low polymerization temperatures are preferable during non-covalent imprinting to protect template-monomer binding (76, 88).

Most commercial monomers contain polymerization inhibitor to avoid undesired polymerization during storage and must be purified by distillation prior to use (76). In addition, the dissolved oxygen present in the polymerization mixture prior can slow down the polymerization process and can be removed by ultrasonication or by Ar or N_2 purging (70). Finally, it is necessary to remove the template after polymerization to free the binding sites. This is often achieved by solvent washing or extraction. Typically, 95-100% of the template can be removed by solvent washing (71).

Characterization and testing

Imprinted polymers are characterized by their morphology (particle size, porosity and surface area) and their specificity in terms of molecular recognition behavior. The most common methods for analyzing morphology include the dry state nitrogen sorption and mercury intrusion porosimetry. Scanning electron microscopy (SEM) is usually employed to observe the porous structures of materials.

The evaluation of the imprinting efficiency (i.e. whether or not the imprinted polymer adequately and accurately memorizes the template) is traditionally done by equilibrium binding assays or chromatographic experiments. The degree of imprinting effect is evaluated by comparing the activity of the imprinted polymer to that of the non-imprinted polymer (NIP) synthesized under the same conditions but without adding the template. Both methods allow values of binding and selectivity to be calculated. However, comparisons of binding capacity should be made between data acquired using the same technique, since chromatographic and equilibrium binding assays do not appear to give the same quantitative results (89).

For equilibrium binding assays, a series of a known amounts of MIPs and NIPs are incubated with different concentrations of the template for a specified time necessary to reach equilibrium. The solution is separated from the polymer by filtration or centrifugation, and the concentration of template in the liquid phase is determined by various analytical methods such as HPLC, UV-spectroscopy, fluorescence or radioactivity (90-92). Generally, the polymers exhibit binding through both specific and non-specific interactions. The non-specific interactions are assessed by measuring the binding of NIPs, while the specific interactions originate from the recognition sites created during the imprinting procedure (93).

Chromatographic evaluation (in columns packed with MIPs used as the stationary phase) compares the retention factor of the template on the MIP with that of the NIP, or the retention of the template with that of other compounds. If the imprinted polymer is efficient, the retention factor of the template on the imprinted phase will be high. If the imprinting effect is poor, the retention factor of the template will be similar to those of other compounds (70). The ratio between the values of the retention factors defines the imprinting factor (94). Although equilibrium binding assays are more informative, chromatographic procedures are straightforward, and precise data are fairly easy to obtain (70). Furthermore, the degree of affinity of structural analogues can be studied (95).

Whereas the methods described above are conducted in organic solvents, environmental applications of MIPs for pollutant removal involves aqueous phases. The concentrations used in traditional testing are also much higher that the level found in water resources. Because many EDCs (and in particular E2) are highly hydrophobic, one could expect a much higher degree of non-specific binding in water. In addition, specific recognition is based on non-covalent molecular interaction which can be influenced by the solvent. For these reasons, it might be more relevant for environmental applications to directly assess the efficiency of the polymers in aqueous phase at low pollutant concentration. For this purpose, solid phase extraction (SPE) can be conveniently used. The typical SPE sequence includes 1) packing of SPE columns/cartridges with known amount of MIP or NIP; 2) solvent washing (if necessary); 3) activation of the sorbent by passing an appropriate solvent (conditioning, if necessary); 4) sample elution; and 5) elution of the target pollutant with a specific solvent (96).

Efficiency of removal of EDCs from water samples using imprinted polymers

There is now a large amount of experience and knowledge available on the synthesis of selective imprinted polymers using hormonally active templates. However, nearly all of these studies have aimed to develop polymers for analytical applications and emphasis was given on high selectivity in organic solvent. For that reason, even when the polymers are tested in SPE, a washing step is normally added after sample elution and before pollutant recovery to remove non-specifically bound interfering compounds and enhance the specific binding of the target molecules. In other words, it is generally believed non-specific hydrophobic interactions prevail in aqueous medium, especially for hydrophobic EDCs such as E2.

For water treatment purpose however, emphasis should be given on highly efficient removal at trace concentrations. The use of imprinted polymers only becomes advantageous if pollutant removal is not affected by the presence of interfering compounds (which is rendered possible by the selective binding sites) and if the adsorbing material can remove unknown compounds having the same biological effects. Hence, the binding of interfering substances is irrelevant as long as pollutant removal is not affected by it but high removal efficiency at environmentally relevant concentrations is of paramout importance given the potential chronic toxicity of EDCs. With this regards, only very few studies have focused on the use of MIPs for pollutant control in a relevant manner.

Removal of EDCs at trace concentration

Table 4 summarizes the findings from various studies aiming to produce MIP for the removal of E2 from aqueous samples. Meng et al. (86) adopted the precipitation polymerization protocol to synthesize micropsheres with high selectivity for α-oestradiol used as template. The efficiency of the imprinted microsphere to remove E2 from spiked lake water was also demonstrated in a batch assay. However, both tests were conducted using a very high pollutant concentration (> 25 mg/L) and in the presence of surfactant. Using a similar protocol, Zhongbo and Hu (85) synthesized E2-imprinted microspheres and demonstrated their selectivity. However, their focus being to study to adsorption isotherm of oestrogens, no tests were conducted at relevant concentration. Le Noir et al. (97) used the traditional bulk polymerization technique to produce E2-MIP with an adsorption capacity 10% higher than that of their NIP. The polymers were then tested for the removal of E2 from pure water spiked with 1 μg E2/L (1 ppb). E2 was recovered 98 ± 2% when using the MIP, compared to 90 ± 1, 79 ± 1, and 84 ± 2% when using the NIP, a commercial C18 phase, or granulated activated carbon, respectively. Assuming the entire fraction of E2 adsorbed was extracted, the concentrations of E2 in the treated samples were below 0.02 ppb for the MIP compared to 0.1 - 0.2 ppb for the other phases. This means the MIP was capable of producing an effluent quality 5–10 times higher than the other adsorbents tested. The higher efficiency of the MIP suggested that specific binding occurred in aqueous phase. Following this study, Le Noir and co-authors synthesized various MIP using variations of the same protocol and achieved similar results in terms of removal efficiencies and selectivities (Table 3).

Influence of interferences and treatment of real water samples.

Le Noir et al. (96) demonstrated the specific advantage of molecular recognition adsorption in the presence of fluoxetine hydrochloride (FH) and acenaphthene (Ac) as interfering compounds. E2 recovery by the MIP (95.5 ± 4.0%) did not decrease significantly in the presence of Ac and FH whereas the NIP efficiency dropped to 54.5 ± 9.4% from 77 ± 5.2% initially, in the absence of Ac and FH. Larger amounts of FH and Ac were recovered from the NIP than from the MIP. E2 recovery from the C18 and GAC phases also decreased. In a more recent study evaluating imprinted microspheres to remove phenolic estrogens, Lin et al. (101) found the polymer efficiency did not decrease in the presence of humic acid. By comparison, a decrease in E2 adsorption capacity to activated carbon has been reported in the presence of surfactant and humic acid (48, 102).

Molecular imprinted polymers have also been tested with real water samples. As above mentioned, Meng et al. (86) achieved E2 removal at high concentration in spiked lake water with 1% Triton 100 as surfactant. This was more recently confirmed by Lin et al. (101) when testing various MIPs in spiked tap, lake or river water although these results were also achieved at very high pollutant concentrations (0.5 mM). Le Noir et al. (96) tested an E2 imprinted polymer for its ability to remove unidentified estrogenic activity from a wastewater sample. Estrogenic activity was only significant in the extracts from the MIP column percolated with wastewater to a level equivalent to the effect caused by 22 ± 4 ng E2/L in wastewater. These results demonstrated imprinted polymers can serve to reduce the total estrogenicity of water streams. Unfortunately, only approx. 100 mL of wastewater could be extracted through each column before clogging. This problem was later solved by embedding the polymers into high-flow gel (98, see Chapt xx for further discussion).

Table 4: Characteristics and properties of MIPs synthesized with E2 as template and tested for water purification.

Protocol	Size (µm)	Pore volume (cm³/g)	BET surface areas (m²/g)	Testing	Specificity/Removal Efficiency	Ref.
Bulk polymerization: E2 (0.272 g), 4VP (0.420 mL), EDMA (5mL), AIBN (50 mg), chloroform (7.5 mL), 65°C for 20 h	≤25			Isotherm in toluene; SPE (100 mg); 1 ppb E2 in water	10% adsorption capacity difference MIP/NIP; 98± 2% with MIP/ 90± 1 % with NIP	97
Bulk polymerization: E2 (0.272 g), 4VP (0.430 mL), EDMA (4.72 mL), AIBN (50 mg), acetonitrile (8 mL), 65°C for 20 h	≤25	MIP (1.04) NIP (1.08)	MIP (335) NIP (367)	Isotherm in toluene; SPE (100 mg); 2 ppb E2 in water	10% adsorption capacity difference MIP/NIP; 100±0.6% MIP/ 77±5.2% NIP	96
Bulk polymerization: E2 (1 mmol), 4VP (4 mmol), EDMA (25 mmol), AIBN (50 mg), acetonitrile (8 mL), 65°C for 20 h	38-106	MIP (0.94) NIP (0.76)	MIP (376) NIP (318)	Isotherm in toluene	10% adsorption capacity difference MIP/NIP	98
Bulk polymerization: E2 (272.4 mg), MAA (0.68 mL), EDMA (4.7 mL), AIBN (100 mg), acetonitrile (8 mL), 43°C for 20 h.	38-106			SPE (100 mg); 2 ppb E2 in water	73±11% MIP/ 46±13% NIP	99
Bulk polymerization: E2 (272 mg), MAA (0.68 mL), EDMA (4.72 mL), 4,4'-azobis(4-cyanovaleric acod) (100 mg), acetonitrile (8 mL), 55°C for 20 h.	38-150	MIP (0.07) NIP (0.07)	MIP (170) NIP (159)	SPE (200 mg); 5 ppb E2 in water	MIP (65.5%) - NIP (46.7)	100
Precipitation polymerization: E2 (50.0 mg), MAA (75.0 mg), TRIM (254.4 mg), 4,4'-azobis(4-cyanovaleric acod) (2 mg), acetonitrile (10 mL), UV235 24 h at 4°C.	0.2 µm microspheres			Isotherm in acetonitrile; Isotherm in water with E2 at 1 ppm (4 µM)	20% adsorption capacity difference MIP/NIP; Similar efficiencies. Approx 10% in difference in MIP/NIP adsorption capacities at 2.9 µM	86
Precipitation polymerization: E2 (0.5 mmol), acrylamide (3 mmol), TRIM (3 mmol), 2,2'-azobis-(2,4-dimethylvaleronitrile) 0.035 mmol), acetonitrile (40 mL), UV360, 15 hr, 4°C	1-2 µm microspheres			Isotherm in aqueous solution containing 1% Triton 100 (vol). estradiol at 0.27 g/L (0.1 mM) concentration.	60% adsorption capacity difference MIP/NIP	85

Regeneration and reuse

Alternatives to imprinted polymers as selective adsorbent include biomolecules such as antibodies and receptor proteins. However, the use of biomolecules to trap endocrine disruptors may be less favorable because proteins are prone to decay. By comparison, imprinted polymers exhibit high physical and chemical resistance against mechanical stress, high temperature and pressure, are often resistant against treatment with acid and base, and are stable in a wide range of solvents. Molecularly imprinted materials can be reused more than 100 times and stored for over 6 years without loss of their "memory effect" (102, 103).

All studies described in Table 4 reported that imprinted polymers could be easily regenerated by solvent washing under normal conditions of temperature and pressure and reused several times without loss of efficiency (without poisoning of the specific binding site). An important limitation to the use of GAC remains is difficult regeneration and tendency to saturate (48, 62). Economical recovery of polymers however needs to be demonstrated at a large scale if microspheres are to be used in suspension.

Conclusions and future recommendations

The potential of molecular imprinted polymers for EDCs removal has been demonstrated in various configurations, at trace concentration, in the presence of interfering compounds, and from real aqueous samples (lake, river, tap and waste waters). Molecular recognition-based adsorption uses the same property that makes the pollutants so harmful: their ability to bind with natural receptors. The use of adsorbents with receptor-like properties should therefore allow the removal of even undetected and unknown pollutants of endocrine disrupting activity, which was indirectly demonstrated through the removal of unknown EDCs from wastewater However, for that potential to be exploited, it is necessary to design MIPs with group-selectivity as it would virtually be impossible to synthesize a MIP to remove each potential pollutant. (96). Meng et al. (86) also showed the MIP synthesized exhibited a significant binding affinity for estrogens other than the template and was therefore appropriate for treating mixtures of estrogenic pollutants. In another study, Ikegami et al (104) showed an imprinted polymer synthesized with the endocrine disruptor bisphenol A as template yielded a polymer with good binding towards the template as well as weak, but clear binding of steroid molecules such as oestradiol. Here was thus an artificial receptor being produced capable to operate on several pollutants of similar molecular structure and toxic properties. This critical property must be further investigated.

As second important area for future research concerns the scaling-up of the process. Currently, micrometer size polymer particles are packed in columns, through which water is passed, or suspended in the aqueous medium and later recovered by centrifugation. Because of the small size and hydrophobic nature of the polymers, the first configuration requires the water stream to be forced through the imprinted material packed in columns that are prone to clogging. Likewise, recovery of micro-size suspended polymers by centrifugation or

20

membrane filtration (unpublished data) is technically difficult or economically prohibitive for water treatment applications. New alternatives must therefore be found. Finally, there is a need to study the long term stability (potential template and precursor leaching) of the polymers.

References

1. Townsend, M. *The Observer* August 8, **2004**.
2. Daugthon, C. *The Lancet* **2002**, *360*, 1035-1036.
3. Daughton, C.; Ternes, T. A. *Environ. Health Perspect.* **1999**, *107*, 907-938.
4. Halling-Sørensen, B., Nielsen, S. N., Lanzky, P. F.; Ingerslev, F.; Lützhøft, H. C. H.; Jørgensen, S. E. *Chemosphere* **1998**, *36*, 357-393.
5. Birkett, J. W.; Lester, J. N. *Endocrine Disruptors in Wastewater and Sludge Treatment Processes*; Lewis Publishers: London, **2003**.
6. Colborn, T.; Dumanoski, D.; Peterson, M. J. *Our Stolen Future: are we threatening our fertility, intelligence, and survival? : a scientific detective story*; Dutton Books: New York, **1996**.
7. Desbrow, C.; Routledge, E. J.; Brighty, G. C.; Sumpter, J. P.; Waldock, M. *Environ. Sci. Technol.* **1998**, *32*, 1549-1558.
8. Baronti, C.; Curini, R.; D'Ascenzo, G.; Di Corcia, A.; Gentili, A.; Samperi, R. *Environ. Sci. Technol.* **2000**, *34*, 5059-5066.
9. Nasu, M.; Goto, M.; Kato, H.; Oshima, Y.; Tanaka, H. *Water Sci. Technol.* **2001**, *43*,101-108.
10. Cargouet, M.; Perdiz, D.; Mouatassim-Souali, A.; Tamisier-Karolak, S; Levi, Y. *Sci. Total Environ.* **2004**, *324*, 55-66.
11. Larsson, D. G. J.; Adolfsson-Erici, M.; Parkkonen, J.; Pettersson, M.; Berg, A. H.; Olsson, P. E.; Forlin, L. *Aquat. Toxicol.* **1999**, *45*, 91-97.
12 Ternes, T. A.; Stumpf, M.; Mueller, J.; Haberer, K.; Wilken, R.D.; Servos, M. *Sci. Total Environ.* **1999**, *225*, 81-90.
13. Huang, C. H.; Sedlak, D. L. *Environ. Toxicol. Chem.* **2001**, *20*, 133-139.
14. Isobe, T.; Shiraishi, H.; Yasuda, M.; Shinoda, A.; Suzuki, H.; Morita, M. *J. Chromatogr. A* **2003**, *984*, 195-202.
15. Lagana, A.; Bacaloni, A.; De Leva, I.; Faberi, A.; Fago, G.; Marino, A. *Anal. Chim. Acta* **2004**, *501*, 79-88.
16. Xiao, X. Y.; McCalley, D.V.; McEvoy, J. *J. Chromatogr. A* **2001**, *923*, 195-204.
17. Kuch, H. M.; Ballschmiter, K. *Environ. Sci. Technol.* **2001**, *35*, 3201-3206.
18. Johnson, A. C.; Belfroid, A.; Di Corcia, A. *Sci. Total Environ.* **2000**, *256*, 163-173.
19. Fujita, M.; Ike, M.; Mori, K.; Kaku, H.; Sakaguchi, Y.; Asano, M.; Maki, H.; Nishihara, T. *Water Sci. Technol.* **2000**, *42*, 23-30.
20. Farre, M.; Kloter, G.; Petrovic, M.; Alonso, M. C.; de Alda, M. J. L.; Barcelo, D. *Anal. Chim. Acta* **2002**, *456*, 19-30.
21. Sekela, M.; Brewer, R.; Moyle, G.; Tuominen, T. *Water Sci. Technol.* **1999**, *39*, 217-220.
22. Komori, K.; Okayasu, Y.; Yasojima, M.; Suzuki, Y.;Tanaka, H. *Water Sci. Technol.* **2006**, *53*, 27-33.

23. Loos, R.; Hanke, G.; Umlauf, G.; Eisenreich, S.J. *Chemosphere* **2007**, *66*, 690-699.
24. Nakada, N.; Tanishima, T.; Shinohara, H.; Kiri, K.; Takada, H. *Water Res.* **2006**, *40*, 3297-3303.
25. Gomez, M. J.; Martinez Bueno, M. J.; Lacorte, S.; Fernandez-Alba, A. R.; Aguera, A. *Chemosphere* **2007**, *66*, 993-1003.
26. Fernandez, M. P.; Ikonomou, M. G.; Buchanan, I. *Sci. Total Environ.* **2007**, *373*, 250-269.
27. Vethaak, A. D.; Lahr, J.; Schrap, S. M.; Belfroid, A. C.; Rijs, G. B. J.; Gerritsen, A.; de Boer, J.; Bulder, A. S.; Grinwis, G. C. M.; Kuiper, R. V.; Legler, J.; Murk, T. A. J.; Peijnenburg, W.; Verhaar, H. J. M.; de Voogt, P. *Chemosphere* **2005**, *59*, 511-524.
28. Kolpin, D. W.; Furlong, E. T.; Meyer, M. T.; Thurman, E. M.; Zaugg, S. D.; Barber, L. B.; Buxton, H. T. *Environ. Sci. Technol.* **2002**, *36*, 1202-1211.
29. Kim, S. D.; Cho, J.; Kim, I. S.; Vanderford, B. J.; Snyder, S. A. *Water Res.* **2007**, *41*, 1013-1021.
30. Heberer, T. *Toxicol. Lett.* **2002**, *131*, 5-17.
31. Kimura, K.; Hara, H.; Watanabe, Y. *Environ. Sci. Technol.* **2007**, *41*, 3708-3714.
32. Yu, J. T.; Bouwer, E. J.; Coelhan, M. *Agr. Water Manage.* **2006**, *86*, 72-80.
33. Jiries, A. G.; Al Nasir, F. M.; Beese, F. *Water, Air, Soil Pollut.* **2002**, *133*, 97-107.
34. Singh, K. P.; Mohan, D.; Sinha, S.; Dalwani, R. *Chemosphere* **2004**, *55*, 227-255.
35. Karvelas, M.; Katsoyiannis, A.; Samara, C. *Chemosphere* **2003**, *53*, 1201-1210.
36. Busetti, F.; Badoer, S.; Cuomo, M.; Rubino, B.; Traverso, P. *Ind. Eng. Chem. Res.* **2005**, *44*, 9264-9272.
37. Ternes, T. A. *Water Res.* **1998**, *32*, 3245-3260.
38. http://ec.europa.eu/research/endocrine/projects_clusters_en.html
39. http://www.epa.gov/scipoly/oscpendo/
40. http://www.epa.gov/ppcp/
41. Bendz, D.; Paxéus, N. A.; Ginn, T. R.; Loge, F. J. *J. Hazard. Mater.* **2005**, *122*, 195-204.
42. Morris, S.; Lester, J. N. *Water Res.* **1994**, *28*, 1553-1561.
43. Ahel, M.; Giger, W.; Koch, M. *Water Res.* **1994**, *28*, 1131-1142.
44. Meakins, N. C.; Bubb, J. M.; Lester, J. N. *Chemosphere* **1994**, *28*, 1611-1622.
45. Carballa, M.; Omil, F.; Lema, J. M.; Llompart, M.; García-Jares, C.; Rodréguez, I.; Gómez, M.; Ternes, T. *Water Res.* **2004**, *38*, 2918-2926.
46. Carballa, M.; Omil, F.; Lema, J. M. *Water Res.* **2005**, *39*, 4790-4796.
47. Frimmel, F. H.; Zwiener, C. *Water Res.* **2000**, *34*, 1881-1885.
48. Fukuhara, T.; Iwasaki, S.; Kawashima, M.; Shinohara, O.; Abe, I. *Water Res.* **2006**, *40*, 241-248.
49. Jones, O. A. H.; Green, P. G.; Voulvoulis, N.; Lester, J. N. *Environ. Sci. Technol.*, **2007**, *41*, 5085-5089.
50. Petrovíc, M.; Gonzalez, S.; Barceló, D. *Trends Anal. Chem.* **2003**, *22*, 685-696.

22

51. Zwiener, C.; Frimmel, F. H. *Sci. Total Environ.* **2003**, *309*, 201-211.
52. Guieysse, B.; Hort, C.; Platel, V.; Munoz, R.; Ondarts, M.; Revah, S. *Biotechnol. Adv.* **2008**, *26*, 398-410.
53. Auriol, M.; Filali-Meknassi, Y.; Tyagi, R. D.; Adams, C. D. *Water Res.* **2007**, *41*, 3281-3288.
54. Junghanns, C.; Moeder, M.; Krauss, G.; Martin, C.; Schlosser, D. *Microbiol.* **2005**, *151*, 45–57.
55. Kim, Y.-H.; Seo, H.-S.; Min, J.; Kim, Y.-C.; Ban, Y.-H.; Han, K. Y. ; Park, J.-S.; Bae, K.-D.; Gu, M. B.; Lee, J. *J. Appl. Microbiol.* **2006**, *102*, 221-228.
56. Lee, S. M.; Lee, J. W.; Park, K. R.; Hong, E. J.; Jeung, E. B.; Kim, M. K.; Kang, H. Y.; Choi, I. G.; *J. Environ. Sci. Health., Part B* **2006**, *41*, 385–397.
57. Suzuki, K;. Hirai, H.; Murata, H.; Nishida, T. *Water Res.* **2003**, *37*, 1972–1975.
58. Soares, A.; Jonasson, K.; Terrazas, E.; Guieysse, B.; Mattiasson, B. *Appl. Microbiol. Biotechnol.* **2005**, *66*, 719–725.
59. Tamagawa, Y.; Hirai, H.; Kawai, S.; Nishida, T. *FEMS Microbiol. Lett.* **2005**, *244*, 93–98.
60. Blánquez, P.; Guieysse, B. *J. Hazard. Mater.* **2008**, *150*, 459-462.
61. Zhang, Y.; Zhou, J. L. *Water Res.* **2005**, *39*, 3991-4003
62. Tchobanoglous, G.; Burton, F.L.; Stensel, H. D. *Wastewater Engineering Treatment and Reuse 4th ed*; McGraw-Hill: New York. **2003**.
63. Ternes, T. A.; Meisenheimer, M.; McDowell, D.; Sacher, F.; Brauch, H. J.; Haist-Gulde, B.; Preuss, G.; Wilme, U.; Zulei-Seibert, N. *Environ. Sci. Technol.* **2002**, *36*, 3855-3863.
64. Ternes, T. A.; Stüber, J.; Hermann, N.; Mdowell, D.; Ried, A.; Kampmann, M.; Teiser, B. *Water Res.* **2003**, *37*, 1976-1982
65. Yoon, Y.; Westerhoff, O.; Snyder, S. A.; Wert, E. C. *J. Memb. Sci.* **2006**, *270*, 88-100.
66. Wintgens, T.; Gallenkemper, M.; Melin, T. *Desalination* **2002**, *146*, 387-391.
67. Yano, K.; Karube, I. *Trends Anal. Chem.* **1999**, *8*, 199-204.
68. Haginaka, J. *Bioseparation* **2001**, *10*, 337-351.
69. Andersson, L. I. *Bioseparation* **2002**, *10*, 353-364.
70. Komiyama, M.; Takeuchi, T.; Mukawa, T.; Asanuma, H. *Molecular Imprinting*; WILEY-VCH Verlag GmbH & Co. KGaA: Weinheim, Germany, **2003.**
71. Mayes, A. G.; Whitcombe, M. *J. Adv. Drug. Deliver. Rev.* **2005**, *57,* 1742-1778.
72. Ramstrom, O.; Ansell, R. J. *Chirality* **1998**, *10,* 195-209.
73. Sellergren, B. *Abstracts of Papers of the American Chemical Society* **1997,** 213: 97-IEC.
74. Jobling, S.; Nolan, M.; Tyler, C. R.; Brighty, G.; Sumpter, J. P. *Environ. Sci. Technol.* **1998**, *32,* 2498-2506.
75. Ying, G.-G.; Kookana, R. S.; Ru, Y.-J. *Environ. Int.* **2002**, *28,* 545-551.
76. Cormack, P. A. G.; Elorza, A. Z. *J. Chromatogr. B.* **2004**, *804,* 173-182.
77. Martin-Esteban, A. *Fresenius. J. Anal. Chem.* **2001**, *370,* 795-802.
78. Yilmaz, E.; Mosbach, K.; Haupt, K. *Anal. Sci.* **1999**, *36,* 167-170.

79. Haupt, K. *Analyst.* **2001,** *126,* 747-756.
80. Andersson, L. I.; Paprica, A.; Arvidsson, T. *Chromatographia.* **1997,** *46,* 57-62.
81. Zhang, H.; Ye, L.; Mosbach, K. *J. Mol. Recognit.* **2006,** *19,* 248-259.
82. Wang, J.; Guo, R.; Chen, J.; Zhang, Q.; Liang, X. *Anal. Chim. Acta* **2005,** *540,* 307-315.
83. Alexander, C.; Andersson, H. S.; Andersson, L. I.; Ansell, R. J.; Kirsch, N.; Nicholls, I. A.; O'Mahony, J.; Whitcombe, M. J. *J. Mol. Recognit.* **2006,** *19,* 106-180.
84. Sellergren, B.; Shea, K. J. *J. Chromatogr.* **1993,** *635,* 31-49.
85. Zhongbo, Z.; Hu, J. *Water Res.* **2008,** *42,* 4101-4108.
86. Meng, Z.; Chen, W.; Mulchandani, A. *Environ. Sci. Technol.* **2005,** *39,* 8958-8962.
87. Andersson, L. I.; Muller, R.; Vlatakis, G.; Mosbach, K. *Proc. Natl. Acad. Sci. U. S. A.* **1995,** *92,* 4788-4792.
88. Spivak, D. A. *Adv. Drug. Deliver. Rev.* **2005,** *57,* 1779-1794.
89. Garcia-Calzon, J. A.; Diaz-Garcia, M. E. *Sensor. Actuat. B-Chem.* **2007,** *123,* 1180-1194.
90. Batra, D.; Shea, K. J. *Curr. Opin. Chem. Biol.* **2003,** *7,* 434-442.
91. Lulka, M. F.; Iqbal, S. S.; Chambers, J. P.; Valdes, E. R.; Thompson, R. G.; Goode, M. T.; Valdes, J. J. *Mater. Sci. Eng.* **2000,** *11,* 101-105.
92. Ye, L.; Weiss, R.; Mosbach, K. *Macromolecules* **2000,** *33,* 8239-8245.
93. Shi, X. Z.; Wu, A. B.; Qu, G. R.; Li, R. X.; Zhang, D. B. *Biomaterials* **2007,** *28,* 3741-3749.
94. Pichon, V. *J. Chromatogr. A.* **2007,** *1152,* 41-53.
95. Zurutuza, A.; Bayoudh, S.; Cormack, P. A. G.; Dambies, L.; Deere, J.; Bischoff, R.; Sherrington, D. C. *Anal. Chim. Acta* **2005,** *542,* 14-19.
96. Le Noir, M.; Lepeuple, A.-S.; Guieysse, B.; Mattiasson, B. *Water Res.* **2007,** *41,* 2825-2831.
97. Le Noir, M.; Guieysse, B.; Mattiasson, B. *Water Sci. Technol.* **2006,** *53,* 205-212.
98. Le Noir, M.; Plieva, F.; Hey, T.; Guieysse, B.; Mattiasson, B. *J. Chromatogr. A* **2007,** *1154,* 2 b-164
99. Fernández-Álvarez, P.; Le Noir, M.; Guieysse, B. *J. Hazard. Mater.* **2009,** DOI: 10.1016/j.jhazmat.2008.07.085
100. Guieysse, B.; Le Noir, M.; Mattiasson, B. *Proceedings of the 235th ACS National Meeting,* New Orleans, LA, April 6-10, **2008.**
101. Lin, Y.; Shi, Y.; Jiang, M.; Jin, Y.; Peng, Y.; Lu, B.; Dai, K. *Environ. Pollut.* **2008,** *153,* 483-491.
102. Mosbach, K.; Ramstrom, O. *Bio/Technology* **1996,** *14,* 163.
103. Vlatakis, G.; Andersson, L. I.; Muller, R.; Mosbach, K. *Nature* **1993,** *361,* 645-647.
104. Ikegami, T.; Mukawa, T.; Nariai, H.; Takeuchi, T. *Analytica Chimica Acta* **2004,** *504,* 131-135.

Chapter 3

Removal of Endocrine Disrupting Contaminants from Water Using Macroporous Molecularly Imprinted Selective Media

Mathieu Le Noir [1,2], Fatima Plieva [1,3], Bo Mattiasson [1*]

[1] Department of Biotechnology, Lund University, P.O. Box 124, SE-22100 Lund, Sweden
[2] Anoxkaldnes AB, Klosterängsvägen 11A, 22647 Lund, Sweden
[3] Protista Biotechnology AB, IDEON, SE-22370 Lund, Sweden

During recent decades we have witnessed growing scientific concern, public debate and media attention concerning the possible adverse effects on humans and wildlife that may result from exposure to a wide range of environmental contaminants. These contaminants are called endocrine disrupters, or endocrine-disrupting compounds (EDCs) and are defined as chemicals that interfere with the function of the hormonal system of organisms.

Changes in the normal hormonal functions in wildlife is not new – some plants produce natural oestrogenic compounds called phytooestrogens – but this man-made effect is being regarded with increasing interest. The scientific community has begun to compile a growing list of man-made chemicals (xenobiotics) that have been shown in laboratory studies to have the ability to interact with and modify the endocrine system. Many of the compounds shown to have endocrine-disrupting effects are commonly used in a large variety of applications, and may therefore be widely dispersed throughout the environment. In fact, EDCs include some pesticides, bulk-produced chemicals, flame retardants, agents used as plasticizers, cosmetic ingredients, pharmaceuticals, natural products such as plant-derived oestrogens, and many more.

The presence of EDCs in the environment has been correlated with a number of reproductive disorders both in humans and wildlife. In wildlife these can include abnormal development of genitalia, imposex (females developing male reproductive organs), intersex (the presence of both male and female reproductive organs), uneven sex ratios (relative numbers of males and females in a population), a decline in reproductive success and an increase in birth defects. In humans, these substances have been implicated in the decline of sperm counts, in cancers of the reproductive organs and in the early onset of puberty.

This is a serious problem as these compounds are used in large quantities in everyday life, and their effects are potentially severe, even at trace concentrations (ng/L-μg/L). Up to 80 man-made chemicals have been found simultaneously in North American sewage, surface water and even groundwater at trace concentrations. Furthermore, while the list of chemicals suspected of being EDCs is long, it is probably still far from complete. The presence of these compounds in the environment thus reflects the fact that conventional wastewater treatment methods are inefficient for their removal.

After 40 years of investigations and analyses, only 12 persistent organic pollutants, including highly toxic dioxins, PCBs, and pesticides such as DDT have been banned by the Stockholm convention (2001). Do we have time to wait several years for more pollutants to be banned? But also, can we afford to carry out the same extensive investigations of each pollutant? Since the number of suspected EDCs is so large, it is not economically feasible to use highly protective mechanisms, such as banning the use of suspected EDCs, to prevent them from entering the environment. Also, economic analysis indicates that treating wastewater with advanced techniques may be economically and environmentally undesirable due to the increased energy consumption and associated economic costs, as well as CO_2 emission. There is thus an urgent need to develop novel, specific and highly effective methods to enrich EDCs in water treatment processes prior to their removal and destruction.

Adsorption processes are widely used in wastewater treatment with the purpose of "polishing" water that has already recieved normal biological treatment. Adsorbents based on granular activated carbon (GAC) are commonly used in advanced wastewater treatment for the removal of organic contanminants due to the avaliability and lower cost of the GAC compared to e.g. silica-based adsorbents (C_{18}) and synthetic polymers) (1). In fact, GAC is able to adsorb pesticides, PAHs and PCBs (1), but also oestrogens (2, 3). However, GAC is difficult to regenerate, requiring high pressure and/or temperature, and also tends to saturate (2, 4). The lack of specificity of previous methods considerably reduces their efficiency at trace concentrations as most of the adsorbent is wasted in the removal of other, often harmless, compounds. Therefore, the use of molecularly imprinted polymers (MIPs) under different types and process configurations is a promising method for the removal of endocrine disrupters at trace concentrations (2, 3, 5-10).

Molecular imprinting is a technique used to induce molecular recognition properties in synthetic polymers in response to the presence of a target molecule acting as a template during the formation of the three-dimensional structure of the polymer (6). Highly efficient and selective adsorption of EDCs from aqueous solutions was recently achieved using molecularly imprinted polymers (MIP) (2, 9). Despite the excellent efficiency of MIPs concerning capture of EDCs at trace concentration, clogging and back-pressure problems when handling particulate containing fluids such as wastewater effluents seriously limit their application. Another important issue is the possibility of processing large volumes of wastewater effluents at high flow rates. The development of the novel, cost-effective sorbents allowing for the fast processing of particulate containing fluids is therefore a major challenge in areas as environmental pollution.

Cryotropic gelation (cryogelation) is an advanced technique that permits formation of macroporous polymeric materials (known as *cryogels*) with controlled porosities. *Cryogels* (from the Greek κριοσ (kryos) meaning frost or ice) are synthesized in semi-frozen aqueous media where ice crystals act as porogen and form continuous interconnected pores after melting. One of the most attractive features of cryogels is their macroporosity, sufficient enough for processing complex fluids containing suspended particles (with average sizes up to 10 μm) *(11, 12)*. The cryotropic gelation renders it possible to form macroporous cryogels filled with solid particles (filler particles) to form so-called composite cryogels. When MIP particles (prepared using different templates) were used as filler, macroporous composite cryogels with high selectivity to the specific targets were prepared *(5, 13, 14)*. The basic concept for the preparation of MIP particles and inherent features for the preparation of the composite cryogel media and the composite highly selective macroporous media are discussed in this chapter.

Synthesis and Selectivity of Molecularly Imprinted Polymers (MIPs)

In modern molecular imprinting, two basic approaches can be distinguished based on the interactions used in the imprinting step: i) the covalent approach, in which the complexes in solution prior to polymerization are maintained by (reversible) covalent bonds, and ii) the non-covalent approach, in which the pre-polymerization arrangement of the template and the functional monomers is formed by non-covalent interactions *(6)*.

The current upsurge in interest in the technique can be attributed to the development of the conceptually and practically simple non-covalent approach. This "self-assembly" approach is similar to natural processes, since most biomolecular interactions are non-covalent in nature *(15, 16, 17)*. Furthermore, the method is quite simple, easy to perform and control, and is by far the most widespread in the literature *(18)*.

The non-covalent approach relies on the formation of a pre-polymerization complex between a monomer carrying suitable functional groups and the template through non-covalent bonds, such as ionic interactions or hydrogen bonding. Using this approach, the functional groups are held in position by the polymeric network. Following polymerization, the template can be removed by solvent extraction. The principal means of re-binding the target molecule to these polymers is achieved again through non-covalent interactions *(6)*. A schematic representation of the non-covalent approach is shown in Figure 1. Designing and synthesizing a MIP can be a challenge because of the number of experimental variables involved, e.g. the nature and proportions of the template, functional monomer, cross-linker, solvent and initiator, and the polymerization method *(19)*. 17β-estradiol (E2) was chosen as the template for being considered as the most active estrogen and for being frequently detected in wastewater. The MIP was prepared by dissolving E2 in acetonitrile in a dried 30 mL test tube, adding the functional monomer, the cross-linker and the initiator and gently mixing the solution for 5 min (Table I). The mixture was then cooled on ice and purged with nitrogen during 5 min before the test tube was sealed and the mixture heated at 65°C for 20 h. Then, the polymer monolith was withdrawn and crushed. The particles were ground in a mechanical mortar in order to obtain fractions of the MIP particles of 38-106 μm and washed with methanol. The flask containing fine particles of polymers in methanol was kept in a fume hood until the methanol was completely evaporated. The dried polymers were finally washed with methanol using a Soxhlet extractor for 24 h and dried in a dessicator for 24 h. To serve as controls, a non-imprinted polymer (NIP) was prepared using the same protocol but without adding the template

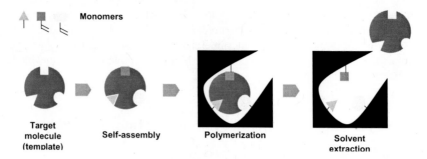

Figure 1. Schematic representation of the non-covalent approach for molecular imprinting.
(see page 1 of color insert)

Table I. Preparation of MIP for 17β-estradiol. Reproduced from *(2)* with permission.

Component	Compound	Quantity	Molar ratio
Template	17β-estradiol	0.272 g	1
Monomer	4-vinylpyridine	0.430 mL	4.05

Various analytical methods can be used to investigate the morphology of imprinted polymers. One of them, the nitrogen sorption porosimetry provides information about the specific surface area (m^2/g), specific pore volume (cm^3/g), average pore diameter (Å) and pore size distribution. A well-defined amount of dry polymer is exposed to nitrogen at a series of pressures. The sorption isotherms are then constructed by measuring the amount of nitrogen sorbed as a function of pressure. The above mentioned information can be extracted from this analytical method obtained from the data of the adsorption-desorption isotherms, following application of BET theory and mathematical models *(20, 21)*. The synthesized MIP and NIP (particles size 38–106 μm) had the specific pore volumes of 0.94 cm^3/g and 0.76 cm^3/g and BET-specific surface areas of 376 m^2/g and 318 m^2/g, respectively and were therefore comparable in their main characteristics (5). The largest volume contributed by either polymer involved pores of between 900 Å and 1100 Å (5).

The evaluation of the imprinting efficiency (i.e. whether or not the imprinted polymer adequately and accurately memorizes the template) is experimentally determined by either chromatographic experiments or equilibrium binding assays. However, equilibrium binding assay is the most commonly used method, and it involves the analysis of imprinted polymers in a solution of substrate (22). Usually, a series of known amounts of MIP and NIP are incubated with different concentrations of the template for a specified time necessary to reach equilibrium. The solution is separated from the polymer by filtration and centrifugation, and the concentration of template in the liquid phase is determined by HPLC using UV, fluorescence or radioactivity detection techniques (23-25). The template selectivity is analysed by comparing this value with the amount of template bound by the polymer with those of other compounds. This approach was used to evaluate the efficiency of the synthesized MIP. The equilibrium binding assay, performed in toluene, showed that the MIP bound 15% more E2 than the NIP (5). Generally, the polymers exhibit binding through both specific and non-specific interactions. The non-specific interactions are assessed by measuring the binding of NIP, while the specific interactions originate from the recognition sites created during the imprinting procedure (26).

However, equilibrium binding assays allow the evaluation of the imprinting efficiency in organic solvents while solid phase extraction allows investigating the specificity of the MIP in aqueous solutions. In a recent book, SPE is defined as, "a method of sample preparation that concentrates and purifies analytes from solution by sorption onto a disposable solid-phase cartridge, followed by elution of the analyte with solvent appropriate for instrumental analysis" (27). Thus, the typical SPE sequence includes sorbent cleaning (if necessary), activation of the sorbent by passing an appropriate solvent through it (conditioning), application of the sample, removal of interfering compounds (clean-up) and elution of the analytes (28). This method was evaluated using MIP synthesized with E2 and NIP and the E2 recovery was found to be 100 ± 1% and 77 ± 3% from the two kinds of polymers, respectively. This suggests that the higher efficiency of the MIP in aqueous phase was due to the specific binding sites.

Cryogelation - Innovative Technology Platform for the Production of Macroporous Polymeric Adsorbents

The cryogelation platform renders it possible to prepare macroporous materials from practically any gel forming systems and with wild range of porosity (12, 29). The cryogels are produced via a gelation process at subzero temperature when most of the solvent is frozen while the dissolved substances (monomers or macromers, cross-linker, initiating system) are concentrated in small non-frozen regions (or non-frozen liquid microphase), where the chemical reaction and gel-formation proceed with time. While all reagents are concentrated in the non-frozen liquid microphase, some part of the solvent remains non-frozen and provides the solutes accumulated into non-frozen part with sufficient molecular or segmental mobility for reactions to perform. An acceleration of chemical reactions performed in non-frozen liquid microphase compared to the chemical reaction in bulk solution is often observed within a defined range of negative temperatures *(30)*. After melting the solvent crystals (ice in case of aqueous media), a system of large continuous pores is formed (Figure 2). Thus the shape and size of the crystals formed determine the shape and size of pores formed after defrosting the sample. The pores size depends on the freezing rate and freezing temperature and initial concentration of monomers/macromers in solution, content of cross-linker, thermal prehistory of the reaction mixture, sample volume, presence of nucleation agents etc. *(31-36)*.

The porosity of cryogels depends often on polymer precursors used. For exemple, using poly(vinyl alcohol) (PVA, one of the most studied synthetic polymers) of different grade (i.e. with degree of hydrolysis from 87-88%) it is possible to form PVA cryogels with pore size up to 1-2 μm *(37, 38)* or with pores up to 100 μm *(35)*. A first type of PVA cryogels was prepared in the shape of beads or membranes *(38 - 41)*, while a second type was often designed as monolithic devices, i.e. monolith columns *(11, 12, 42)*. The typical feature of cryogel monoliths is the continuous system of large interconnected pores with average pore size distribution 10-100 μm. Due to the continuous large pores in the cryogel monoliths, the mass transfer is mainly due to convection *(43)*, resulting in a very low back pressure through the monoliths *(43 - 45)* and scaled-up composed monoliths *(46)*.

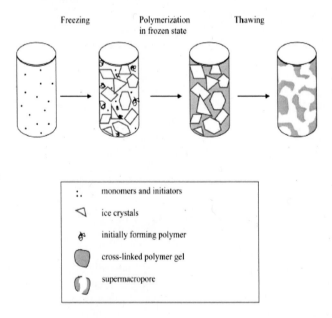

Figure 2. Schematic representation of cryogels preparation. Reproduced from (44) with permission.

There are different techniques to form macroporous adsorbents with desired properties. In most typical case, cryogels with large interconnected pores are prepared via a single freezing-thawing cycle when required ligands or active fillers are incorporated *in situ* during the cryogel synthesis. The multilayered cryogels are prepared via sequential freezing-thawing cycles *(47)*. The sequential freezing is straightforward approach to form macroporous cryogels with controlled porosities and desired gel surface chemistries. Namely, cryogels with double-continuous macroporous networks were prepared via synthesis of new cryogel network inside the interconnected macropores of already formed cryogels *(47)*. The porous structure of these cryogels was shown to consist of two spatially separated and continuously interlacing cryogel networks as was shown by Confocal Scanning Electron Microscopy analysis of cryogels stained with the fluorescent dyes Rhodamine and FITC (Figure 3) (from *(47)* with permission).

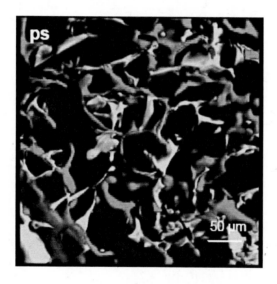

Figure 3. Confocal Laser Scanning Microscopy (CLSM) 3D reconstituted images of macroporous cryogel with double-continuous macroporous networks and stained with fluorescent dyes. The primary PEG-cryogel was stained with Rhodamine B (red colored walls) and the secondary PEG-cryogel was stained with FITC (green colored walls). Reproduced from (47) with permission. (see page 1 of color insert)

The presence of large continuous pores in the macroporous cryogel monoliths give ability to the cryogels to let particulate matter pass, e.g. microbial cells can pass through a bed of cryogels without being retained *(48-51)*. The same time, the large size of pores in macroporous cryogels with total pore volume of 90 % results in low available surface area for binding a specific target. There are different ways to increase the capacity of cryogels for binding with specific targets and, the same time, to keep the macroporous structure of cryogels. These approaches include decreasing (to some extent) the pores size in cryogels *(31, 47, 52)*, grafting of polymer chains on to the surface of the cryogels to increase the binding sites with targets *(53-57)*, preparation of multilayered cryogel systems *(47)* or preparation of composite cryogels *(5, 13, 58, 59)*. The formation of the composite cryogel systems is an excellent approach for the development of macroporous cryogel systems of increased capacity. This approach allowed for many design considerations as differently sized filler particles of required surface chemistry and of different nature (i.e. fillers of organic or inorganic nature) can be used for the preparation of composite systems.

Inherent Features for the Preparation of Macroporous Composite Media

Macroporosity is one of the most attractive features of cryogels. Having pore sizes large enough for the free passage of bioparticles as intact cells, the macroporous cryogel monoliths present an excellent matrix for incorporation of differently sized fillers. It was shown in many examples that incorporated fillers i) improve mechanical strength of macroporous cryogels *(59, 60)*, ii) do not decrease essentially the flow-path properties of macroporous cryogel arranged as monolithic cartridges *(5, 59)* and iii) in some cases improve the re-swelling abilities of cryogels from a dried state *(5, 59)*. Depending on the filler size, the porous structure of the composite cryogels will be different. One could distinguish three main structures: i) the composite cryogels with filler particles incorporated into the polymer walls (when the size of filler particles is less than the thickness of the cryogel walls), ii) a composite cryogels with filler particles partly incorporated into polymer walls and partly exposed into the macropores (when the filler particles with broad particle distribution (e.g. 1-50 μm) are used) and iii) composite cryogels with filler particles simply entrapped into macropores (when the size of filler particles is comparable to or larger than the size of macropores in cryogels).

The presence of filler particles in a reaction mixture influences on the freezing pattern of the reaction mixture and eventually results in the formation of cryogels with uniform porous structure. The reason for that is that the solid particles (filler), distributed in the reaction mixture, perform as centers of ice nucleation and facilitate ice crystallization. It was shown that presence of filler particles (nanosized magnetic particles or microsized titania particles) in the reaction mixture resulted in diminishing the temperature of overcooling of the reaction mixture (overcooling or supercooling is a phenomenon when water is cooled below the freezing point (0 °C) without ice nucleation and remains liquid to temperatures as low as -40 °C) *(59)*.

Solid particles of both inorganic and organic nature can be used as filler. Inorganic fillers as surfactant stabilized Fe_3O_4 nanoparticles, silica, hydroxyapatite and titania microparticles were used for the preparation of high flow-through macroporous composite cryogel monoliths *(58, 59, 61)*. As fillers of organic nature, polymeric microparticles synthesized from both synthetic and natural precursors can be used. Thus, the commercially available Amberlite (cross-linked polystyrene) *(52)*, Sephadex (cross-linked dextran) *(60)* or MIP particles (cross-linked with ethylene glycol dimethacrylate 4-vinylpyridine or methylmethacrylate) *(5,13)* were used for the preparation of composite macroporous cryogels in beaded shape *(52, 60)*, as high flow-through composite cryogel monoliths *(5)* or as composite cryogels in protective shells *(13)*.

Fillers in a wide range of sizes, from nanosized (58, 59) till microsized particles (5, 52, 59, 61) were used for the preparation of composite cryogel systems and evaluation of their properties. Typically, the content of filler in the composite systems is varied in the range of 5-50 % (not published data). The content of filler depends on the potential application of cryogel composite systems. The accessibility of functional ligands/specific binding sites in the incorporated filler for binding with specific targets is one of the most important parameters to be taken into account when designing the composite cryogel systems with active fillers.

Preparation of Composite Cryogels from Different Gel Precursors

Composite cryogels prepared via cryogelation technique can be divided on two main groups: thermo-reversible physically cross-linked composite cryogels (where the cryogel network is formed through hydrogen bond formation) and thermo-irreversible covalently cross-linked cryogels (when the cryogel network is formed through covalent bond formation). The physically cross-linked composite cryogels were prepared through physical gelation at subzero temperatures of both natural polymers like agarose or synthetic polymer like PVA *(52, 60)*. Covalently cross-linked composite cryogels are produced via two main approaches, namely chemical cross-linking using an appropriate cross-linkers or through the free radical cross-linking polymerization. According to the first approach, the cross-linking reaction is performed in partially frozen media using the available cross-linkers as, for example glutaraldehyde (GA) and epichlorohydrin (ECH). Thus, chitosan-based *(5)* and PVA-based composite cryogels *(5)* were prepared via cross-linking of PVA and chitosan chains with GA in acidic medium. According to the second approach, the covalently cross-linked cryogels were produced through free-radical cross-linking polymerization using ammonium persulfate (APS) and N,N,N',N'-tetra-methyl-ethylenediamine (TEMED) initiating system *(5, 59, 62)*.

Formats of Cryogels and Composite Cryogels

The cryogelation platform renders it possible to form macroporous cryogel and composite cryogel systems of different formats and sizes. Basically, any format for the plain cryogels (i.e. without filler particles) is applicable for the formation of composite cryogel systems. The most used formats for cryogel and composite cryogel systems are beads, films/sheets, monoliths/discs and cryogels formed in protective shells (Figure 4). The geometrical format of cryogel is to some extent caused by porosity of the cryogels and depends on particular application. Thus, composite cryogels with pore sizes up to 1-2 µm (like PVA cryogels prepared from PVA with a degree of hydrolysis more than 97 % *(37)*) were prepared mainly in the shape of beads or films *(52, 60)*, while the macroporous cryogels with pore sizes up to 100 μm were produced in the shape of beads (size more than 1 mm), sheets, monoliths/discs or they were arranged inside a protective plastic shell or housing *(5, 13, 14, 58, 59, 63)*. In the later case (cryogels in protective shells), the open-ended plastic housings of different format and sizes can be used.

Beads   Sheets

Monoliths (discs)

Monoliths in protective shells (MGPs)

Figure 4. Different formats of macroporous cryogels: beads, sheets, monolithic rods or discs and cryogels-in-protective-shells.

Macroporous Composite Selective Media

The combination of high flow-through pores in cryogel monoliths with high density of functional groups/binding sites on the microporous filler allows for the preparation of macroporous composite systems with increased capacities on binding of specific targets. Thus, the highly selective macroporous composite system was prepared via combining high selectivity of MIP with high flow-through properties of macroporous cryogel monoliths *(5)*. The cryogel prepared from monomer precursors (as acrylamide) and polymer precursors (as chitosan and poly(vinyl alcohol)) were used for the preparation of the selective macroporous media *(5)*. The choice of the gel precursors for the preparation of the cryogel systems was caused be their availability and robustness. The concentration of gel precursors and ratio of cryogel to MIP particles was optimized to prepare the control cryogels and composite cryogel monoliths with high flow-path properties and pore volume more than 85% (Table II). It was shown, that incorporation of up to 25 vol% of MIP particles into the reaction mixture for the preparation of cryogels, resulted in the preparation of the composite MIP/cryogel cartridges with high flow-path abilities (Table II). The incorporation of MIP particles did change the flow-through properties of the macroporous cryogels (even though resulted in slight decreasing the flow-through properties compared to plain cryogels) and, for some cryogel systems, resulted in improving the swelling abilities from the dried state (Table II). The incorporation of the MIP particles to the reaction mixture resulted in preparation of the more mechanically stable monolith systems. The compressibility of the chitosan cryogels (which are compressible by itself up to 70 % under water flow) decreased till 18-20 % due to the incorporation of the solid MIP particles (Table II).

The MIP/PVA columns and the MIP columns (of similar volume) were characterized with very low back pressure (Figure 5). Noteworthy, that the back pressure for the composite MIP/PVA and MIP columns were much lower compared to the commercially available Sepharose-4B (agarose based matrices) (Figure 5). SEM analysis showed the microporosity of MIP with cavities/channels of less than 1 µm in size (Figure 6a). When combined with macroporous network of a cryogel, the distribution of the MIP particles in the final composite cryogel network depended much on the size of MIP particles. Thus, MIP particles with particle sizes up to 25 µm were mainly incorporated into cryogel walls (Figure 6B), while MIP particles of larger sizes (38-106 µm) were partly incorporated into the walls of the cryogel (with exposed surfaces of MIP particles into pores) or were mechanically entrapped into macropores of the cryogel (Figure 6C). It is worth noticing that leakage of MIP particles from the composite cryogel media was minimal thereby making it possible for repeated use of the composite cryogel system for extraction of the specific targets.

When preparing the composite cryogel system for the selective capture of specific targets, it was important to ensure that there was no non-specific binding to the cryogel network. It was shown that it was no non-specific binding of E2 to PVA cryogel as no extraction peak was obtained by HPLC analysis compared to the extraction peaks obtained for the PVA/MIP and PVA/NIP systems (Figure 7). The higher efficiency was obtained for the MIP/PVA system compared to NIP/PVA system (i.e. without the specific sites for binding) (Figure 7). The recovery of E2 from PVA/MIP (104 ± 9%) was significantly higher compared to PVA/NIP (49 ± 4%) and other analyzed earlier commercial sorbents as C18 (79 ± 1%) and GAC (84 ± 2%) *(2)* (Figure 7).

Table II. Porous properties of cryogels and composite cryogel/MIP monoliths. Reproduced from (5) with permission.

Gel precursors used for MG preparation	Type of monolith	Notation	Pore volume [a] (%)	Water flow path through the monolith column [b] (cm/h)	Re-swelling after drying (% of initial weight), time
PVA (5.4 %)	control	PVA	88.6	620 [c]	(100), within 1-2 min
	composite	PVA/MIP	83.6	580	(100), within seconds
pAAm (7 %)	control	pAAm	90.2	820	(100), within seconds
	composite	pAAm/MIP	86.4	770	(100), within seconds
Chitosan (0.8 %)	control	Chs	Not checked	1100 [c]	(16-22), h
	composite	Chs/MIP		835	(70-75), within 1-2 min

[a] Pore volume was estimated as described elsewhere (31)
[b] The water flow path through the monolith column was estimated at the same hydrostatic pressure equal to 1 m water column
[c] The PVA and chitosan cryogel monoliths were compressed up to 25 and 70 % of their initial height, respectively

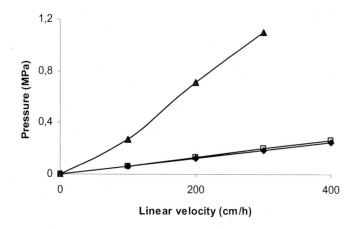

Figure 5. The pressure drop for composite MIP/PVA column (closed diamonds) and MIP column (open squares) of similar volume (inner diameter 12 mm, volume 2 mL). The pressure drop for the column of similar size packed with Sepharose CL-4B (open triangles) is presented for comparison. The pressure drop experiments were performed using a Bio-Rad FPLC system using deionized water as a mobile phase. The water was passed through the columns for 1 min at each flow rate and pressure was recorded.

The large, μm-in-size interconnected pores in the PVA/MIP composite allowed for fast processing of wastewater without clogging, while columns packed with MIP itself were rapidly clogged due to the retained particulate matter (Table III). The real wastewater was passed through the composite PVA/MIP columns at a flow rate as high as 50 mL min^{-1} while it was possible to pass about 55 mL of the particulate containing fluid through the MIP column at a flow rate of 1 mL min^{-1} (Table III). Due to the accumulated particulate matter, the water flow rate through the MIP column was decreased from 1 mL min^{-1} till about 0.2 mL min^{-1} until complete blocking the column. The application of E2 solution to the composite PVA/MIP columns and extraction of E2 were performed at a high flow rate of 50 mL min^{-1} which corresponds to the linear velocity of 2500 cm h^{-1}. The processing at a high flow rate was possible due to the large size of interconnected pores in the composite PVA/MIP systems. Also, the efficiency of capturing E2 from the applied solution did not decrease at high flow rates. More than 90 % of the captured E2 was possible to extract from the composite PVA/MIP system with 4 mL of the elution solvent (or two column volumes). Two liters of a solution of E2 was applied to the PVA/MIP monolith columns and E2 was extracted within 40 min compared to 6 h required for processing the same volume of the E2 solution using the MIP column at a flow rate of 5 mL min^{-1}.

38

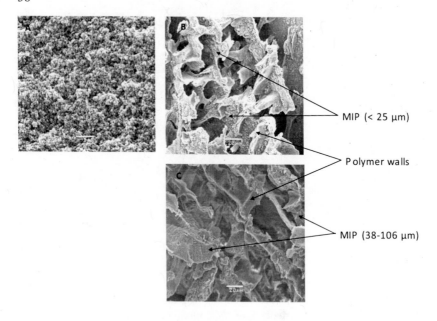

MIP (< 25 µm)

Polymer walls

MIP (38-106 µm)

Figure 6. Scanning Electron Microscopy (SEM) images of MIP (A) and composite MIP/PVA cryogel (B, C) prepared using the MIP particles of different sizes: <25 µm (B) and 38-106 µm (C). Images (A) and (C) reproduced from (5) with permission.

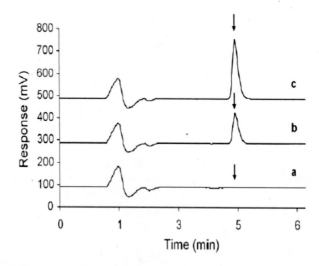

Figure 7. Extraction chromatogram obtained for the PVA (a), PVA/NIP (b) and PVA/MIP (c) columns. Experimental conditions: 2 L of an aqueous solution containing 2 µg L^{-1} of 17β-estradiol were applied to PVA, PVA/NIP and PVA/MIP columns at a flow rate of 50 mL min^{-1}. Extraction was performed with methanol: acetic acid (4:1 v/v) solution. Arrows show the elution peak of E2 through the monolith columns. Reproduced from (5) with permission.

It is important to note that the composite PVA/MIP columns were used for the removal of E2 from water repeatedly. The sorbent medium was easily regenerated with more than 99 % E2 recovery and showed the same efficiency on repetitive sorption-extraction experiments *(5)*. High flow path abilities of the prepared composites allowed for processing and regeneration at high flow rates which made the experiments using the composite cryogels/MIP systems very fast. The ease of regeneration is one of the important features of the prepared PVA/MIP composites compared to the granulated activated carbon, which is difficult to regenerate and tends to saturate *(1, 4)*. To demonstrate the importance of specific binding under real conditions, E2 was applied in the presence of acenaphthene (Ac) and humic acid (HA) as interfering substances that can be found simultaneously as E2. Ac has a similar hydrophobicity to that of E2 (Kow ≈ 4); it is considered as a weak-estrogen that should not bind to estrogen receptors *(64)*. HA was selected to simulate dissolved organic matter as the E2 adsorption capacity of GAC (granulated activated carbon) can drop by a 1000-fold in river water due to site-poisoning and/or site-competition *(65)*. Under

Table III. Application of wastewater effluent to the MIP, PVA cryogels and composite PVA/MIP cryogel columns. Reproduced from *(5)* with permission.

Material	MIP 100 mg	PVA 1.6 mL	MIP/PVA 100 mg/1.6 mL
Applied wastewater volume (mL)	55	200	200
Flow of wastewater applied (mL min⁻¹)	1 [a]	50	50
Recovery (%) [b]	10	100	100

[a] After passing wastewater sample for 180 minutes, the MIP column was completely clogged.

[b] Recovery (%) was estimated as follows: (Absorbance of wastewater effluent at 420 nm after passing the column / Absorbance of wastewater effluent at 420 nm before passing the column) x 100

these conditions, E2 recoveries dropped to 54.4 ± 7.7 % and 29.5 ± 7.5 % for the PVA/MIP and PVA/NIP system, respectively (Figure 8), but the selectivity on the imprinted material was still maintained. Ac bound much less to the PVA/MIP (4.4 ± 0.5%) than to the PVA/NIP (15.5 ± 3.3 %) (Figure 8), suggesting the presence of specific sites favoured the adsorption of the target molecule and reduced the binding of interfering agents. It is often believed that non specific adsorption mechanisms prevail in aqueous phase because of the hydrophobic nature of the polymers. This means that solvent washing is required to remove non-specifically-bound interferences and strengthen, or leave unaffected, the specifically-bound substances *(66)*. The obtained results showed that specific binding to the molecular site occurred in the aqueous phase, allowing for the higher removal efficiency of the target pollutant with less binding of interfering compounds. This is very advantageous for the treatment of large liquid volumes.

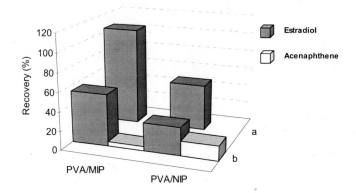

Figure 8. Recovery of 17β-estradiol and acenaphthene from PVA/MIP (black bars) and PVA/NIP (white bars). Experimental conditions: 2 L of an aqueous solution containing 2 μg L^{-1} of 17β-estradiol (a) or 2 L of an aqueous solution of 17β-estradiol, acenaphthene and humic acid at concentration of 2μg L^{-1} each (b) were applied to the composite monoliths at a flow rate of 50 mL min^{-1}. The 17 β- estradiol was extracted from the composite monoliths with methanol:acetic acid (4:1 v/v) solution scale applications as compared to granulated activated carbon. Reproduced from (5) with permission.

Macroporous Composite Selective Media in Protective Shells

The recent design of macroporous cryogels in protective shells allowed for expanding the potential applications of macroporous cryogels to mechanically stirred processes *(14, 67)*. The formation of cryogels (with up to 100 μm-sized pores) inside an open-ended protective housing was shown a versatile approach to protect these highly porous materials from abrasion caused by collisions due to the stirring *(14, 67)*. Thus the macroporous composite media prepared using different templates were prepared inside protective housings. For these purposes the composite MIP/PVA and NIP/PVA cryogels were formed inside so-called Kaldnes carriers (Figure 4, cryogels-in-protective-shells). These plastic carriers are of special design and shaped like a cylinder with a length 7 mm and a diameter 10 mm (K1 type) or with length of 10 mm and a diameter of 25 mm (K3 type) (Figure 4). These carriers are used in a suspended carrier process where biomass grows on the plastic carriers that are suspended in the reactor because of the agitation caused by aeration (in aerobic reactors) or by mechanical mixing (in anaerobic reactors*) (68, 69)*. These carriers are robust and available and represent an appropriate "housing" for the cryogels.

The swollen hydrophilic MIP/PVA cryogels were prepared inside the K1-type Kaldnes carriers. The composite PVA/MIP cryogels occupied completely the housing space with no voids between gel and plastic wall. Both

PVA (control) and composite MIP/PVA MGPs (prepared using K1-carriers) were stable to attrition and withstood intensive stirring for prolonged time (data not shown). It was shown that the composite MIP/PVA MGPs completely removed E2 from applied solution at stirring at 200 rpm and there was some non-specific adsorption of E2 to control PVA MGPs, while there was no absorption of E2 to the plastic carriers themselves *(14)*. Thus, it was shown that the MGPs bearing selectivity to a specific template can be used for the removal of trace contaminants from water effluents under stirring.

One of the unique features of MGPs is the possibility of using them in different column configurations as for columns packed with MGPs or in the columns with fluidized bed reactors. These both types of reactor configurations were evaluated for the capture of EDC from water effluents at environmentally relevant concentrations *(13)*. The MIP/PVA composite media prepared using E2,

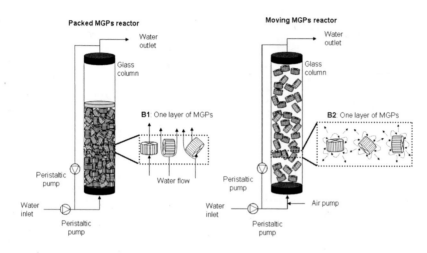

Figure 9. Schematic representation of a packed-bed MGP reactor (left) and a moving-bed MGP reactor (right). B1: Schematic representation of one layer of MGP in the packed-bed reactor. The arrows show the direction of water flow through the column. B2: Schematic representation of one layer of MGP in the moving-bed reactor. The arrows show the motion of the MGP in the column.

atrazine and 4-nonylphenol (NP) as the templates were evaluated in packed-bed reactor and moving-bed reactor (Figure 9). Comparison of performance of the different reactor configurations showed that the removal of E2 contaminant was more efficient in the moving-bed configuration than in the packed-bed configuration. Furthermore, the binding of E2 by the selective adsorption medium (E2-MIP/MGP) was independent of the flow rate, and was the same for flows of 1 mL min^{-1} (which corresponded to a HRT of 120 min) and 15 mL min^{-1} (which corresponded to a HRT of 8 min). Complete binding of E2 was achieved after 400 minutes of recycling the E2 solution through the column at both flow rates in the moving-bed reactor (Figure 10). However, the binding of E2 was strongly dependent on the flow rate in the packed-bed reactor (Figure 10). Only 70% of the E2 was removed by the selective adsorption medium in the packed-bed reactor at a flow rate of 1 mL min⁻¹ (corresponding to a HRT of 120 min), and the corresponding value at a higher flow rate of 15 mL min⁻¹ (corresponding to a HRT of 8 min) was 32%. It was assumed that aeration ensured mixing of the MGP and better accessibility for the target contaminants. Even if the plastic carriers used for the preparation of MGP presented open ends protective carrier

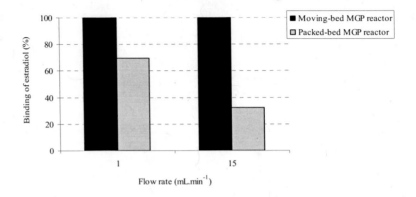

Figure 10. Amount of 17β-estradiol (E2) bound to E2-MIP/MGP in a moving-bed MGP reactor and a packed-bed MGP reactor at flow rates of 1 and 15 mL min⁻¹.

Experimental conditions: 200 mL of an aqueous solution spiked with E2 (1 mg L⁻¹) was pumped through a column filled with the MGP at a flow rate of 1 mL min⁻¹, and one L of aqueous solution spiked with E2 (1 mg L⁻¹) was pumped through a column filled with the MGP at a flow rate of 15 mL min⁻¹. In all cases, the solution was recycled through the columns until no further adsorption of E2 to MGP was detected.

(total empty volume of about 0.44 mL), some part of the macroporous selective medium formed inside the carriers (total volume of macroporous gel embedded with MIP formed inside the Anoxkaldnes carrier is about 0.4 mL) was not assessable for binding at high applied flow rates in the packed MGP reactor configuration. The MGP were packed in a random manner, meaning that the configuration of the MGP was not optimal for processing in the packed-bed reactor, especially at the higher flow rate (inset B1, Figure 9). The performance of the moving-bed MGP reactor allowed more efficient adsorption due to the continuous motion of the MGPs (inset B2, Figure 9), resulting in increased binding of contaminants to the MGP compared to the situation inpacked-bed mode. One could assume that more optimal housing design (e.g. presence of mesh in housing shell) would improve the performance of MGPs in the packed – bed mode.

It was shown that the binding capacities of the selective adsorption medium (E2-MIP/MGP) and the non-selective adsorption medium (E2-NIP/MGP) were 674 nmol mg⁻¹ (at 90% saturation) and 598 nmol mg⁻¹ (at 75% saturation), respectively *(13)*. High specificity of the imprinted adsorption medium for capture of the specific targets was maintained at doubled flow rates. Thus, the kinetic adsorption of E2 by the selective adsorption medium (E2-MIP/MGP) was essentially the same at high flow rates of 15 and 30 mL min^{-1} and showed over 90% binding of the E2 after 4 h of recycling E2 solution in the moving-bed reactor *(13)*. Similar adsorption behaviour was observed for atrazine at flow rates of 15 and 30 mL min^{-1}, with over 80% binding of atrazine after 4 h of recycling the atrazine solution through the selective atrazine-MIP adsorption medium. It was shown that the binding efficiency of E2 and atrazine to their respective selective adsorption medium was essentially maintained when the hydraulic retention time was decreased by up to 30 times (from 120 min for 1 mL min^{-1} to 4 min for 30 mL min^{-1}) (Figure 11).

In order to determine if there was any non-specific binding to the empty Kaldnes carriers or plain PVA-based MGP (*i.e.* with no MIP or NIP particles), the adsorption of E2 and atrazine to the empty carriers and plain PVA-based MGP was studied in the moving-bed reactor. When applying one litre of E2 or atrazine solutions (1 mg L^{-1}) at a flow rate of 15 mL L^{-1} (HRT = 8 min) for at least 6 h of recycling, neither E2 nor atrazine were adsorbed onto the empty plastic Kaldnes carriers (Figure 12). However, under the same conditions, 50% of the E2 and 19% of the atrazine had adsorbed onto the plain PVA-based MGP (Figure 5). This observation can be explained by the different physical and chemical properties of E2 and atrazine *(13)*. All the E2 (100%) was bound by the column containing the E2-MIP/MGP after 6 hours of recycling the same aqueous solution spiked with E2 (Figure 12). This was much higher than the amount of

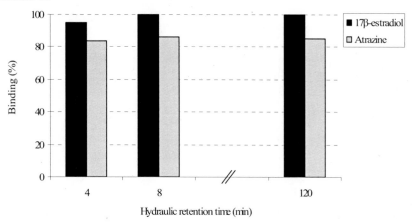

Figure 11. Amount of 17β-estradiol and atrazine bound to E2-MIP/MGP and atrazine-MIP/MGP, respectively, using different Hydraulic Retention Times (HRT) of a solution of 1 mg L^{-1} 17β-estradiol or atrazine in a moving-bed MGP reactor. Samples (1 mL) were taken every 200, 60 and 30 min at HRT of 120, 8 and 4 min, respectively. The values given in this figure are based on the amounts recorded when no further adsorption was observed. The concentration of the contaminants was quantified by HPLC.

E2 bound to the column containing the E2-NIP/MGP (77%), confirming the greater efficiency of the MIP over the NIP, even at high flow rates. On the other hand, 86% of the atrazine was bound to atrazine-MIP/MGP after the same recycling time, while only 52% was bound to a column containing atrazine-NIP/MGP (Figure 12). These results showed that atrazine-MIP/MGP contained specific binding sites for atrazine. However, over 90% E2 binding and about 80% atrazine binding were achieved after passing the spiked solutions (1 mg L^{-1}) through the E2-MIP/MGP and atrazine-MIP/MGP for 4 h, respectively. The binding increased and reached 100% for E2 and 86% for atrazine after 6 hours of recycling through the respective adsorption media (Figure 12). The difference in binding of E2 and atrazine can be related to the different octanol-water partition coefficient (log K_{ow}) of E2 and atrazine (4.01 and 2.82, respectively) *(13)*. It is known that the value of log K_{ow} below 2.5 shows a low sorption potential, whereas the value greater than 4 indicates a high sorption potential *(70)*. Nevertheless, 77% of the E2 and 53% of the atrazine were removed by non-imprinted adsorption medium (E2-NIP/MGP and atrazine-NIP/MGP, respectively) after 6 hours of recycling an aqueous solution spiked with 1 mg L^{-1} E2 or atrazine. The difference in capture of E2 and atrazine by the selective(MIP/MGP) and non-selective (NIP/MGP) media showed i) specificity of MIP compared to NIP and ii) less non-specific adsorption for atrazine compared to β-estradiol by non-selective media (atrazine-NIP/MGP compared to E2-NIP/MGP).

The high efficiency and specificity of the imprinted adsorption media was confirmed when studying EDCs at trace concentrations, *i.e.* for the environmentally relevant concentrations of EDCs. All studied EDCs (E2, atrazine and NP) were removed from water at trace concentrations (environmentally relevant concentration) using the imprinted macroporous composite adsorption media in moving-bed reactor configuration.

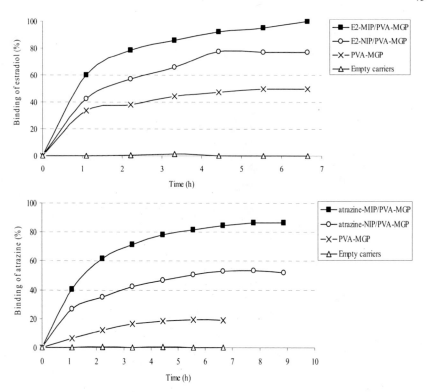

Figure 12. Amount of 17β-estradiol (A) and atrazine (B) bound to different materials (MIP/PVA MGP, NIP/PVA MGP, plain PVA MGP and empty plastic carriers).
Experimental conditions: (A) an aqueous solution (1 L) spiked with E2 (1 mg L^{-1}) was pumped at 15 mL min^{-1} through E2-MIP/MGP, E2-NIP/MGP, plain PVA-based MGP and empty Kaldnes carriers in a moving-bed MGP reactor. This solution was recycled until no further E2 was adsorbed onto the MGP. The content of E2 in the outlet water was quantified over time.
(B) An aqueous solution (1 L) spiked with atrazine (1 mg L^{-1}) was pumped through atrazine-MIP/MGP, atrazine -NIP/MGP, plain PVA-based MGP and empty Kaldnes carriers in a moving-bed MGP reactor. This solution was recycled over time until no atrazine was adsorbed to the adsorption medium. Atrazine was quantified in the water outlet over time.

46

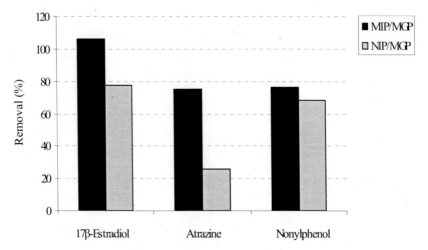

Figure 13. Removal (%) of the three EDCs (17β-estradiol, Atrazine and 4-Nonylphenol) in a moving-bed MGP reactor using molecularly imprinted (MIP) and non-molecularly imprinted (NIP) adsorption media.

Experimental conditions: for 17β-estradiol (E2): 10 L of an aqueous solution spiked with E2 (0.5 μg L⁻¹) was pumped through E2-MIP/MGP at a flow rate of 15 mL min⁻¹. For Atrazine: 5 L of an aqueous solution spiked with atrazine (10 μg L⁻¹) was pumped through atrazine-MIP/MGP at a flow rate of 15 mL min⁻¹. For 4-Nonylphenol (NP): 20 L of an aqueous solution spiked with NP (5 μg L⁻¹) was pumped through NP-MIP/MGP at a flow rate of 15 mL min⁻¹. The captured contaminants were eluted from the MGP column with methanol and the concentrations of E2, atrazine and NP in the eluted fractions were quantified using HPLC.

Efficient removal of all EDCs (E2, atrazine and NP) was obtained from the spiked water using the selective MIP/MGP adsorption media (Figure 13). The behaviour of the selective MIP/MGP media compared to NIP/MGP media was different for all three studied contaminants. Thus, complete removal of E2 was achieved after applying a large volume (10 L) of E2 solution (0.5 μg E2 L⁻¹) to the column filled with the E2-MIP/MGP at a flow rate of 15 mL min⁻¹ (HRT of 8 min), in moving-bed reactor mode, while 77% of the E2 was removed from the applied solution by the non-imprinted medium (Figure 13). These results correlated to those reported for macroporous selective media prepared as monolithic cartridge, when   complete (100 %) removal of E2 was achieved MIP/PVA compared to 72% achieved with non-selective NIP/PVA *(5)*.

Regarding atrazine, 75 and 26% removal was achieved with a HRT of 8 min after applying 5 L of a 10 μg L⁻¹ atrazine solution to the atrazine-MIP/MGP and NIP/MGP, respectively (Figure 13) *(13)*. The removal achieved with atrazine-MIP/MGP was much higher than that observed in the study by Meakins et al *(71)* during conventional secondary treatment (40%). It was shown that the final concentration of atrazine after treatment with atrazine-MIP/MGP medium was 1.22 μg L⁻¹, which is much lower than the current allowed level for atrazine in drinking water (3 μg L⁻¹) *(72)*.

Practically the same efficiency of capturing of NP from the spiked solution was achieved by the selective and non-selective adsorption media (76 and 69% for NP-MIP/MGP and NP-NIP/MGP, respectively) *(13)* (Figure 13). Such a small difference between the selective and non-selective adsorption media was explained by the nature of the template used. NP is hydrophobic and has only one functional group, and its log K_{ow} value (4.48) is even higher than that of E2 *(13)*. Due to its high adsorption potential, NP was bound more non-specifically to its imprinted adsorption medium in contrast to the adsorption media prepared with other templates (E2 and atrazine). Similar data (the same efficiency of binding the NP to the selective MIP and non-selective NIP SPE cartridges) were reported by Soares et. al. *(73)*. Thus it was shown that binding of the specific targets (templates) to the selective MIP/MGP as well as to the non-selective NIP/MGP media depended on the nature of the used template.

Concluding Remarks and Future Perspectives

The composite cryogel adsorbents hold a great promise as robust and highly efficient sorption media for removal of a broad range of contaminants from water effluents at trace concentrations. Molecular imprinting is a useful technique for the preparation of polymeric materials as specific molecular recognition receptors (molecularly imprinted polymers, MIP), which selectively and strongly bind the target molecules (templates) and enable the selective removal of the templates from a mixture of closely related compounds. Besides, MIPs are stablet, cost effective and can be produced in large amounts.

The cryogelation technique offers a possibility for the preparation of macroporous adsorbents with controlled porosities (pore sizes up to 200 μm) and of different formats from a variety of monomer/polymer precursors. The unique combination of properties of macroporous cryogels together with their inertness, chemical and mechanical robustness opens wide perspectives for designing macroporous adsorbent materials with superior performance for different sorption processes including processing the particulate containing fluids.

Combination of both advanced techniques (molecular imprinting and cryogelation) allowed for the preparation of new type of macroporous highly selective materials which can be considered as a new generation of macroporous adsorbents . By embedding the MIP particles into a cryogel network, the composite cryogel monolithic cartridges with high selectivity for capturing the target contaminants (templates) can be prepared. One of the important features of these cartridges is the possibility for processing large volumes of water without sacrificing high efficiency on removal of the target molecule (template) from water (as was shown for the selective removal of β-Estradiol by MIP/cryogel cartridge *(5)*). In fact, any other available adsorbents, widely used for capturing of contaminants (as activated carbons, polystyrene-based beads, etc) can be incorporated into the cryogel networks to form high flow-through composite cryogel cartridges. The specificity of these high flow-through cartridges on the removal of contaminants will vary from non-specific capture of a wide range of contaminants from water (for e.g. active carbons/cryogel system) till highly selective capture of the target contaminant molecule (template) by MIP/cryogel system.

Important issue when designing the composite MIP/cryogel systems is the retension of molecular recognition sites and high selectivity of the imprinted material when embedded in a cryogel matrix. The multilayered composite cryogel systems or the composite cryogel with embedded mixed-MIP filler will be promising for the preparation of highly selective cartridges of multifunctional performance, i.e. selective capture of several contaminants (templates). For example, the MIP particles, prepared using various templates can be used in one cryogel composite system (mixed MIP filler) or the MIP synthesized to different templates can be separately introduced into the different cryogel networks via a sequential freezing-thawing approach *(47)*. Such robust multi-capture highly selective cryogel cartridges will have obvious advantages in cleaning-up of water from specific contaminants at trace concentrations compared to other existing adsorbents.

References

1. Tchobanoglous, G.; Burton, F.L.; Stensel, H.D. Wastewater Engineering Treatment and Reuse. McGraw-Hill, New York, 2003, p 1139.
2. Noir, M.L.; Guieysse, B.; Mattiasson, B. *Wat. Sci. Technol.* **2006,** *53*, 205-212.
3. Noir, M.L.; Lepeuple, A.S.; Guieysse, B.; Mattiasson, B. *Wat. Res.* **2007,** *41*, 2825-2831.
4. Fukuhara, T.; Iwasaki, S.; Kawashima, M.; Shiohara, O.; Abe, I. *Wat. Res.* **2006,** *40*, 241-248.
5. Noir, M.L.; Plieva, F.M.; Hey, T.; Guiyesse, B.; Mattiasson, B. *J. Chromatogr. A,* **2007,** *1154*, 158-164.
6. Komiyama M.; Mukawa, T.; Asanuma, H. Molecular imprinting: from fundamentals to applications. Wiley-VCH, 2003.
7. Caro E.; Marc, M.x.; Cormack, P.A.G. *J. Chromatogr. A* **2004,** *1047*, 175-181.
8. Chapuis, F.; Mullot, J.-U.; Pichon, V.; Tuffal, G.; Hennion, M.-C. *J. Chromatogr. A* **2006,** *1135*, 127-134.
9. Meng Z.; Chen W.; Mulchandani, A. *Environ. Sci. Technol.* **2005,** *39*, 8958-8962.
10. Prasad, K.; Prathish, K.P.; Gladis, J.M.; Naidu, G.R.K.; Rao, T.P. *Sensor Actuat. B: Chem.* **2007,** *123*, 65-70.
11. Lozinsky, V.I.; Galaev, I.Y.; Plieva, F.M.; Savina, I.N.; Jungvid, H.; Mattiasson, B. *Trends Biotechnol.* **2003,** *21*, 445-451.
12. Plieva, F.M.; Galaev, I.Y.; Mattiasson, B. *J. Sep Sci.* **2007,** *30,* 1657-1671.
13. Noir, M.L.; Plieva, F.M.; Mattiasson, B. /Unpublished/. **2008**
14. Plieva, F.M.; Mattiasson, B. *Ind. Eng. Chem. Res.* **2008,** *47*, 4131-4141.
15. Ramstrom, O.; Ansell, R.J. *Chirality,* **1998,** *10*, 195-209.
16. Ansell, R.J.; Mosbach, K. *Analyst* **1998,** *123*, 1611-1616.

17. Sellergren, B. Important considerations in the "design" of receptor sites using noncovalent imprinting. Abstracts of papers of the ACS 213, 1997, 97-IEC.

18. Mayes, A.G.; Whitcombe, M.J. *Adv. Drug Deliv. Rev.* **2005**, *57*, 1742-1778.

19. Cormack, P.A.G.; Elorza, A.Z. *J. Chromatogr. B* **2004**, *804*, 173-182.

20. Brunauer, S.; Emmett, P.H.; Teller, E. *J. Am. Chem. Soc.* **1938**, *60*, 309.

21. Groen, J.C.; Peffer, L.A.A.; Perez-Ramirez, J. *Microp. Mesop. Mater.* **2003**, *60*, 1-17.

22. Pichon, V. *J. Chromatogr. A* **2007**, *1152*, 41-53.

23. Batra, D.; Shea, K.J. *Curr. Pin. Chem. Biol.* **2003**, *7*, 434-442.

24. Lulka, M.F.; Iqbal, S.S.; Chambers, J.P.; Valders, E.R.; Thompson, R.G.; Goode, M.T.; Valdes, J.J. *Mater. Sci. Eng.* **2000**, *11*, 101-105.

25. Ye, L.; Weiss, R.; Mosbach, K. *Macromolecules*, **2000**, *33*, 8239-8245.

26. Shi, X.Z.; Wu, A.B.; Qu, G.R.; Li, R.X.; Zhang, D.B. *Biomaterials* **2007**, *28* 3741-3749.

27. Thurman, E.M.; Mills, M.S. Solid-phase extraction: principles and practice. Wiley-VCH, New-York, 1998.

28. Andersson, L.I. *J. Chromatogr. B: Biomed. Appl.* **2006**, *739*, 163-173.

29. Lozinsky, V.I.; Plieva, F.M.; Galaev, I.Yu.; Mattiasson, B. *Bioseparation* **2002**, *10*, 163-188.

30. Lozinsky, V.I. *Russ. Chem. Rev.* **2002**, *71*, 489-511.

31. Plieva, F.M.; Karlsson, M.; Aguilar, M.-R.; Gomez, D.; Mikhalovsky, S.; Galaev, I.Y. *Soft Matter* **2005**, *1*, 303-309.

32. Ozmen, M.M.; Okay, O. *Polymer*, **2004**, *46*, 8119-8127.

33. Plieva, F.; Huiting, X.; Galaev, I.Y.; Bergenståhl, B.; Mattiasson, B. *J. Mater. Chem.* **2006**, *16*, 4065-4073.

34. Galaev, I.Y.; Dainiak, M.B.; Plieva, F.; Mattiasson, B. *Langmuir*, **2007**, *23*, 35-40.

35. Plieva, F.M.; Karlsson, M.; Aguilar, M.-R.; Gomez, D.; Mikhalovsky, S.; Galaev, I.Yu.; Mattiasson, B. *J. Appl. Polym. Sci.* **2006**, *100*, 1057-1066.

36. Perez, P.; Plieva, F.; Gallardo, A.; Roman, J.S.; Aguilar, M.; Morfin, I.; Ehrburger-Dolle, F.; Bley, F.; Mikhalovsky, S.; Galaev, I.Yu.; Mattiasson B. Biomacromolecules **2008**, *9*, 66-74.

37. Lozinsky, V.I. *Russ. Chem. Rev.* **1998**, *67*, 573-586.

38. Lozinsky, V.I.; Plieva, F.M. *Enz. Microb.Techn.* **1998**, *23*, 227-242.

39. Martins, R.F.; Plieva, F.M.; Santos A.; Hatti-Kaul R. *Biotechnol. Lett.* **2003**, *25*,1537-1543.

40. Das-Bradoo S.; Svensson, I.; Santos, J.; Plieva, F.M.; Mattiasson, B.; Hatti-Kaul, R. *J. Biotechnol.* **2004**, *110*, 273-285.

41. Efremenko, E.N.; Spiricheva, O.V.; Veremeenko, D.V.; Baibak, A.V.; Lozinsky, V.I. *J. Chem. Technol. Biotechnol.* **2006**, *81*, 519-522.

42. Dainiak, M.B.; Galaev, I.Yu.; Kumar, A.; Plieva, F.M.; Mattiasson, B. *Adv Biochem Engin/Biotechnol.* **2007**, *106*, 101-127.

50

43. Persson, P.; Baybak, O.; Plieva, F.; Galaev, I.Yu.; Mattiasson, B.; Nilsson, N.; Axelsson, A. *Biotechnol. Bioeng.* **2004**, *88*, 224-.

44. Plieva, F.M.; Andersson, J.; Galaev, I.Y.; Mattiasson, B. *J. Sep. Sci.* **2004**, *27*, 828-836.

45. Hanora, A.; Plieva, F.M.; Hedström, M.; Galaev, I.Yu.; Mattiasson, B. *J. Biotechnol.* **2005**, *118*, 421-433.

46. Plieva, F.M.; De Seta, E.; Galaev, I.Yu.; Mattiasson, B. *Sep. Pur. Tech.* available online March 27, 2008.

47. Plieva, F.M.; Ekström, P.; Galaev, I.Yu.; Mattiasson, B. *Soft Matter* **2008**, DOI: 10.1039/B804105A.

48. Arvidsson, P.; Plieva, F.M.; Savina, I.N.; Lozinsky, V.I.; Fexby, S.; Bulow, L.; Galaev, I.Yu.; Mattiasson, B. *J. Chromatogr. A* **2002**, *977*, 27-38.

49. Plieva, F.M.; Savina, I.N.; Deraz, S.; Andersson, J.; Galaev, I.Yu.; Mattiasson, B. *J. Chromatogr. B* **2004**, *807*, 129-137.

50. Noppe, W.; Plieva, F.M.; Vanhoorelbeke, K.; Deckmyn, H.; Tuncel, M.; Tuncel, A.; Galaev, I.Yu.; Mattiasson, B. *J. Biotechnol.* **2007**, *131*, 293-299.

51. Dainiak, M.B.; Plieva, F.M.; Galaev, I.Yu.; Hatti-Kaul, R.; Mattiasson B. *Biotechol. Progr.* **2005**, *21*, 644-649.

52. Savina, I.N.; Hanora, A.; Plieva, F.M.; Galaev, I.Y.; Mattiasson, B.; Lozinsky, V.I. *J. Appl. Polym. Sci.* **2005**, *95*, 529-538.

53. Savina, I.N.; Galaev, I.Yu.; Mattiasson, B. *Polymer* **2005**, *46*, 9596-9603.

54. Savina, I.N.; Galaev, I.Yu.; Mattiasson, B. *J. Chromatogr. A* **2005**, *1092*, 199-205.

55. Savina, I.N.; Galaev, I.Yu.; Mattiasson, B. *J. Mol. Recogn.* **2006**, *19*, 313-321.

56. Yao, K.; Yun, J.; Shen, S.; Chen, F. *J. Chromatogr. A* **2007**, *1157*, 246-251.

57. Yun, J.; Shen, S.; Chen F.; Yao K. *J. Chromatogr. B* **2007**, *860*, 57-62.

58. Yao, K.; Yun, J.; Shen, S.; Wang L.; He, X.; Yu, X. *J. Chromatogr. A*, **2006**, *1109*, 103-110.

59. Plieva, F.M.; Pignetti, D.; Galaev, I.Yu.; Mattiasson, B. Porous structure of supermacroporous composite cryogels **2008**, *unpublished.*

60. Lozinsky, V.I.; Zubov, A.L.; Titova, E.F.; Rogozhin, S.V. *J. Appl. Polym. Sci.* **1992**, *44*, 1423-1435.

61. Lozinsky, V.I.; Bakeeva, I.V.; Presnyak, E.P.; Damashkaln, L.G.; Zubov, V.P. *J. Appl. Polym. Sci.* **2007**, *105*, 2689-2702.

62. Yao, K.; Shen, S.; Yun, J.; Wang, L.; He, X.; Yu, X. *Chem. Engin. Sci.* **2006**, *61*, 6701-6708.

63. Plieva, F.M.; Oknianska, A.; Degerman, E.; Mattiasson, B. *Biotechnol. J.*, **2008**, DOI 10.1002/biot.20070134.

64. Makoto, K.; Toshifumi, T.; Takashi, M.; Hiroyuki, A. Molecular Imprinting: From Fundamentals to Applications. Wiley-VCH Verlag, Weinheim, 2003.

65. Zhang, Y.; Zhou, J.L. *Wat. Res.* **2005**, *39*, 3991-4003.

66. Andersson, L.I. *Analyst* **2000**, *125*, 1515-1517.

67.   Önnby, L.; Giorgi, C.; Plieva F.M.; Mattiasson, B. **2008,** *unpublished.*

68.   Dalentoft, E.; Thulin P. *Wat. Sci. Tech.* **1997,** *35,* 123-130.

69.   Odegaard, H.; Gisvold, B.; Strickland, J. *Wat. Sci.Tech.* **2000,** *41,* 383-391.

70.   Birkett, J.W.A.L. Endocrine disruptors in wastewater and sludge treatment process. Lewis publishers, London, 2003.

71.   Meakins, N.C.; Bubb, J.M.; Lester J.N. *Chemosphere* **1994,** *28,* 1611-1622.

72.   EPA, National primary drinking water regulations, 816-F-02-013. Office of Water Washington, DC, 2002.

73.   Soares, A.; Guerreiro, A.; Piletska, E.; Mattiasson, B.; Piletsky, S. *Anal. Chim. Acta* **2008,** *612,* 99-104.

Chapter 4

Imprinted Polymers for the Removal of Hydrophilic and Hydrophobic Metal Complexes

Syed Ali Ashraf, Ckarlos Mercado, Anja Mueller[*]

Department of Chemistry, Central Michigan University, Mt. Pleasant, MI 48858

Imprinting polymerization was used to make high capacity resins for heavy metal ions for wastewater remediation. Resins already used in wastewater treatment were being imprinted. 100% removal efficiency for up to 80 ppm of cadmium and mercury ions with only 200 mg of resin was demonstrated with the imprinted resins, but not the non-imprinted reference samples. For hydrophobic complexes, imprinting emulsion polymerization was developed. The effect of ionic vs. metal-ligand forces on imprinting polymerization is being discussed as well.

Introduction

The remediation of wastewater is difficult because it requires the resin (polymeric material) that performs the remediation to be general and specific at the same time: general, because it has to remove a lot of different contaminants that happen to be in that water at that time, and specific for whatever has to be removed, so that the capacity of the resin is high enough for that specific contaminant to be completely removed. At the same time the resin has to do all of that in a flexible manner, because the type and amount of contaminants is going to vary.

A method used for flexible removal of a variety of particles and contaminants is flocculation[1]. Hydrophilic polymers, for example polyacrylamides, are added to the water to remediate, and offer a surface to the contaminants to adhere to. During the process of adhesion, some charges of the polymer will be neutralized, thus reducing the solubility of the polymer. At the same time, some contaminants will bridge different polymers, creating a large network, and reducing the solubility of the whole "floc" (aggregate) just simply because of size. Eventually the aggregates become large and insoluble and dense enough to sediment to the bottom of the basin.

Another flexible method used for remediation is membrane filtration[2]. Filters are used to remove larger particles in prefiltration to very small compounds and gases in nanofiltration. Membranes are also used in special applications such as ion exchange membranes for desalination of seawater. Both of these methods generally work well but are still too inefficient for heavy metal ions[3,4].

Heavy Metal Ions in Water

In the Great Lakes region of the United States cadmium and mercury are the major polluting heavy metal ions. In the west of the country arsenic, chromium and selenium are the major pollutants. Here cadmium, mercury, and arsenic ions will be discussed.

Cadmium is highly toxic and persistent in the environment. Cadmium occurs naturally in zinc, lead, copper, and other ores. Cadmium is used in batteries, pigments, coatings and plating, stabilizers for plastics, and nonferrous and other alloys.[5] Ores serve as one of the principal sources of cadmium in ground and surface waters, especially when in contact with soft, acidic waters. Industrial sources of cadmium in water are, for example, smelting or recycling of nickel-cadmium (NiCad) batteries[6]. Zinc-bearing coals of the central United States also contain cadmium[5]. The corrosion of galvanized plumbing and water main pipe materials can releases cadmium to drinking water, another common source are leaching land fills.

Symptoms of acute cadmium exposure are nausea, vomiting, diarrhea, muscle cramps, sensory disturbances, liver injury, convulsions, shock, and renal failure[6]. Long term effects of cadmium include kidney, liver, bone, and blood cell damage. About 8% of lung cancers may be attributable to cadmium as well.

Mercury is equally toxic and persistent in the environment. It occurs naturally in few ores, but its major source are industrial, such as metal alloys, chemical catalysts, dental fillings, batteries, and electrical switches[7]. Another major source of pollution are coal-fired power plants[7]. Mercury concentrates in fish, shellfish, seals and whales.

In the body, mercury is the most damaging to the brain, especially in developing brains of children and fetuses. Mercury easily crosses both the blood brain barrier and from the mother to the fetus. Mental retardation is the main result; in adults it may be the cause of Parkinson's, Alzheimer's, and Multiple Sclerosis when ingested in low doses over time. Methyl mercury is so toxic that only one dose, even adsorbed across the skin, will lead to death.

Arsenic ions, again, are highly toxic and also persistent in the environment[8]. About one-third of the arsenic in the Earth's atmosphere is of natural origin[9]. Arsenic leaching from minerals is found in groundwater in several parts of the world, for example Bangladesh, India, Taiwan and the Western United States. Mining, smelting of non-ferrous metals and burning of fossil fuels are the major industrial processes resulting in arsenic contamination[9]. Historically, arsenic was also used in pesticides and preservation of timber.

Symptoms of arsenic poisoning include violent stomach pains, vomiting, thirst; hoarseness and difficulty of speech; diarrhea; convulsions and cramps; delirium and death[8]. Long-term, low dose arsenic poisoning can lead to a variety of problems, for example skin cancer.

Remediation of Heavy Metal Ions in Water

Common resins have been proven inefficient for the removal of heavy metal ions from wastewater[3, 4]. Therefore several laboratories have been working on developing specific resins just for the remediation of one or two heavy metal ions. An adsorbent containing carboxylate groups was prepared by grafting tin (IV) oxide gel onto polyacrylamide, which was shown to have a greater selectivity for Pb^{2+} than Hg^{2+} and Cd^{2+} [10]. Heavy metal ion adsorption is dependent on time, concentration, pH, and temperature. Diallyldimethylammonium chloride was grafted onto polyacrylamide for a flocculant specialized for titanium dioxide[11]. The synthesis of this resin was somewhat difficult and costly. A chelating polymer was prepared from glycidyl methacrylate and iminodiacetic acid[12]. This resin was effective for the adsorption of Cr^{3+} and Cu^{2+} at pH 2-6. Again, a complicated synthesis makes this material more expensive.

Waste materials such as hazelnut shell[13] and lignin[14] were attempted as well. Hazelnut shell is used as a biosorbent substrate for the removal of Cd^{+2}, Zn^{+2}, Cr^{3+}, and Cr^{6+} [15]. The sorption characteristics are very pH dependent; maximum removal was observed only at the pH of 2.5-3.5. Lignin was used as an adsorbent for cations such as Cu^{2+}, Zn^{2+}, Pb^{2+}, and U^{2+} [14]. The adsorption of these ions depends on the nature of the lignin, ionic strength, pH, and the buffer used.

There are increasing regulations for industry requiring reductions in the amount of heavy metals released. Also, the removal of heavy metal ions from water would play a significant role in reducing health problems.

In this project a resin is made specific not by synthesizing a new resin, but by imprinting a resin similar to those used in wastewater treatment with a specific heavy metal ion. In addition to the specific adsorption of the heavy metal ion, random adhesion of other impurities in wastewater to the improved coagulant can still occur. Thus this resin would be sufficient for clarification of water as well as the removal of heavy metal ions.

Imprinting Polymerization

Imprinting polymerization involves the preparation of the polymer in the presence of an template (Fig. 1). When the template is removed, a specific void fitting that molecule remains in the polymer. The imprinted polymer will now allow more efficient and specific binding of the template.

Imprinting polymerization was developed for the separation of enantiomers and is currently still one of the few effective method for this application[16]. Usually it is based on non-covalent, weak forces: hydrogen bonding, dipole-dipole interactions, and in this case also ionic forces. For separation and purification processes, these weak forces are maximized by making and using the imprinted resin in organic, non-polar solvents. In that case excellent specificity can be achieved. Also, generally larger molecules achieve better specificity than smaller ones.

In this research the imprinting polymerization is performed in water to avoid shrinking or swelling of the imprint. In water weak forces are even weaker, reducing the specificity of the imprint. Also, the templates are small heavy metal ion complexes, further reducing the specificity. But for remediation applications the capacity of the resin is more important for the effectiveness of the resin than the specificity, if there are always enough binding sites for the templates to bind, even in the presence of a lot of other molecules that can compete. Thus in this project imprinting is used to add additional, and preferential binding sites. Additionally, large surface area will be used to result in a resin with a large access of binding sites, many of them preferential for heavy metal ions.

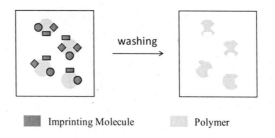

Imprinting Molecule   Polymer

Figure 1. Imprinting polymerization: the polymerization of the resin is performed in the presence of templates. Afterwards the templates are washed out; shapes specific to the templates remain in the resin.

Imprinting Polymerizations inside of Emulsion Polymerizations

Imprinting polymerization is a general method: any molecule can in principle be used as the template, even mixtures of molecules could be used. But it is easier to imprint a hydrophobic polymer with a hydrophobic molecule and a hydrophilic polymer with a hydrophilic molecule. We wanted to extend our method to hydrophobic heavy metal complexes (and in the future other hydrophobic, hard to remediate, impurities) but still perform the polymerization of the - now hydrophobic – polymer resin in water. Therefore, we combined our imprinting technique with an already known method to synthesize hydrophobic polymer in water, emulsion polymerizations[17], to result in an imprinting emulsion polymerization (Fig. 2).   Imprinting   emulsion polymerization has been reported where a core-shell particle is being imprinted on the surface of the particle[18]. The authors were unable to find any previous report of an imprinting polymerization all throughout the monomer micelle in an emulsion polymerization.

In an emulsion polymerization, micelles are formed by a surfactant in water. Hydrophobic monomers aggregate in the inside of those micelles because they prefer the hydrophobic environment. Since only micelles of a certain size are stable, only a certain number of monomers fit inside the micelles. Since each surfactant has their allowed micelle size, one can control the polymer length by the choice of surfactant in the reaction. In fact, emulsion polymerization was developed as a method to control the length of a polymer but is now commonly used because it avoids toxic and expensive solvents.

The best surfactants for stable emulsions, i.e. stable micelles, are ionic, since they repel each other with their charge and thus do not fuse into an oil phase that separates out from the water phase. In this research, though, non-ionic surfactants have to be used, otherwise the heavy metal ions combine with the surfactants and thus do not enter the micelle as the templates. It was found that alkyl-oligo(ethylene glycol) surfactants worked well[19].

With both of these methods it is now possible to imprint resins commonly used in wastewater treatment with any heavy metal ion complex. It will even be possible to imprint with mixtures, if needed. Therefore, one cold take a wastewater stream, concentrate it, and use the resulting solution as the templates – without even having to know the details of the composition. Also, the imprinting method can also be used for different kinds of compounds, such as polyaromatic hydrocarbons (PAH) or other persistent toxic compounds found in water, which are actually easier and more effective templates than heavy metal ion complexes.

58

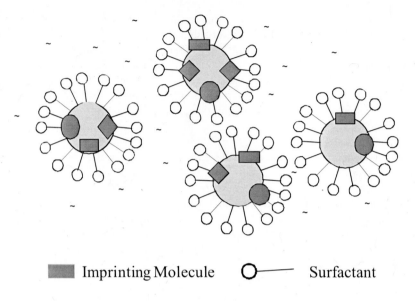

Imprinting Molecule   O—— Surfactant

Figure 2. Imprinting emulsion polymerization: the polymerization of the resin is performed in the presence of templates inside of an emulsion. The hydrophobic templates will travel into the hydrophobic environment within the micelles, where the polymerization is being performed as well. Afterwards the templates are again washed out; shapes specific to the templates remain in the resin.

Materials and Methods

Materials

Methacrylic acid (99%), phenyl methacrylic acid (98%), methacrylamide (98%), acrylamide (99%), 2,2-azobis(2-methylpropionamidine) dihydrochloride (AAPD, 97%), ethylene glycol diacrylate (EDMA, 90%), 1,6-hexanediol diacrylate (80%), potassium persulfate, Brij-56, and mercury(II) dithizonate were obtained from Aldrich Chemical Company. Methyl-methacrylate (99%) was purchased from Acros Organics. Cadmium chloride (99.8%) was obtained from Fisher Chemical Company. Water used in the experiments is purified through Barnstead E-pure filter system and collected at 18 MOhm. Snake skin pleated dialysis tubing, molecular weight cut-off 3,500, was obtained from Pierce.

Polymer Synthesis and Characterization

70:30 Methacrylate:methacrylamide random copolymer samples (1). This is the hydrophilic polymer for the hydrophilic heavy metal ion complexes. For the 70:30 methacrylate:methacrylamide copolymer, 0.70 and 0.30 mol of methacrylate and methacrylamide, respectively, along with 0.001 mol AAPD, 0.0005 mol EDMA, and 150 mL of deionized water were placed in a round-bottom flask under inert nitrogen atmosphere. The mixture was then placed inside a Rayonet photochemical reactor and stirred for eight hours. The polymer obtained was the non-imprinted polymer control. For imprinting polymerization, 0.001 mol of cadmium chloride, cesium chloride, or arsenic trioxide was added to the mixture described above and polymerized in the same manner. To synthesize the porous, imprinted polymer and porous, non-imprinted control, nitrogen gas was purged additionally through the reaction mixture during polymerization; all other conditions and procedures remain the same. After polymerization, all polymers were filtered out and dried. Impurities, unreacted monomer, and cadmium were removed in batches from the imprinted polymers by dialysis in water to result in the washed, imprinted samples. The final yield after dialysis is still being determined. Non-imprinted, non-porous (**1a**) IR: 3600-3000, 2950, 1701, 1458, 1255, 1174, 925, 700, 600, 510 cm^{-1}; imprinted, non-porous (**1b**) IR: 3600-3000, 2950, 1701, 1458, 1255, 1161, 925, 700, 600, 510 cm^{-1}; Non-imprinted, porous (**1c**) IR: 3600-3000, 2950, 1701, 1474, 1388, 1249, 1176, 925, 920, 700, 600, 510 cm^{-1}; imprinted, porous (**1d**) IR: 3600-3000, 2950, 1701, 1482, 1388, 1260, 1178, 925, 920, 782, 510 cm^{-1}.

50:50 Methyl methacrylate:phenyl methacrylate random copolymer samples (2). This is the hydrophobic resin for the hydrophobic heavy metal ion complexes. A round bottom flask was covered with aluminum foil and 3.03 g (0.0044 mol) of Brij-56 surfactant were added followed by 3.85 mL of phenyl methacrylate (0.025 mol) and 2.34 mL of methyl methacrylate (0.0217 mol) and 0.240 g (8.86×10^{-4} mol) of the initiator potassium persulfate under inert atmosphere. 1.12 ml (0.005 mol) of 1,6-hexanediol diacrylate was added prior the addition of the 200 ml of deionized H$_2$O. The reaction was stirred at 70 °C for three hours. To acquire only non-porous product, nitrogen gas was delivered into the flask without introducing it directly into the solution. For imprinting, 0.217g of Mercury (II) dithizonate (3.03×10^{-4} mol) was added last to the reaction. For the porous samples, nitrogen gas was delivered directly into the solution during the reaction. After the reaction was complete, it was cooled down to room temperature. Then 200 ml of water was added to the solution to precipitate the polymer. The solution was placed in the refrigerator for 48 hrs to complete the precipitation after which the polymer was recovered by gravity filtration. The yield from non-porous control (**2a**) product was 2.40 g (38.4%), from non-porous imprinted (**2b**) was 2.25 g (36.5%), from porous control (**2c**) 4.36 g (71.0%), and from porous imprinted (**2d**) 4.65 g (75.7%). All yields are calculated based on the amount of monomer added to the reaction.

Infrared (IR) Spectroscopy. Polymer samples were analyzed with a Nicolet Magna-IR 560 Spectrometer via a KBr pellet.

Differential Scanning Calorimetry (DSC). Samples of 2.00 to 5.00 mg were measured from 10 °C to 80 °C, at a rate of 5 °C per minute using a TA Instruments Inc. model 2100 thermal analysis system equipped with a model 2910 DSC cell. The sample compartment was exposed to a constant purge of dry nitrogen at 50 ml/min.

Thermogravimetric Analysis (TGA). Samples of 5-10 mg in a platinum sample pan were examined by thermogravimetry using a TA Instruments model 2950 TGA unit interfaced with the Thermal analyst model 2100 control unit. TGA was carried out at a heating rate of 10 °C/ min. The TGA cell was swept with nitrogen at 50 ml/min during runs.

BET Surface Area and Pore Size. The surface area and porosity of each polymer were analyzed using Micrometrics ASAP 2020 analyzer. 2.0 g of sample were weighed, degassed for 4 hrs, and then reweighed to ensure that all water was removed. This weight was used as the basis for calculations. The temperature ramp during BET surface area measurements was 1 °C/min, with N2 being the gas that was adsorbed. After analysis the sample was reweighed, and another measurement was taken.

Metal Ion Adsorption Measurements

Column Chromatography. Unless otherwise stated, 200 mg of polymer sample were used as the solid phase in a small chromatography column (I.D. = 0.9 cm). 10 mL of a known concentration of $CdCl_2$, $CsCl$, or AsO_3 solution in water, mercury (II) dithizonate in 80% acetonitrile in water were loaded onto the column and the elute was collected.

Flame Atomic Absorption (AA) Spectroscopy. Cadmium and cesium samples were measured using flame AA with a Perkin Elmer Atomic Absorption Spectrometer 3100. Mercury samples were measured via furnace AA with a Perkin Elmer Zeeman Atomic Absorption Spectrometer 4100ZL.

Furnace AA Spectroscopy. The standard solutions were used to create a standard curve using the instrument Perkin-Elmer AAWinLab for determining the concentrations of samples obtained from Columns. We used a mercury lamp with a current of 6 mA, a wavelength of 253.7 nm, and a slit width of 0.7.

Indirect Measurement of As^{3+} ions via Magnesium Complexation and Flame AA Measurements. To increase the safety of the measurements, arsenic ions were measured indirectly by following the procedure of Hassan and Eldesouky[20]. All arsenate samples were made in 0.1 N HCl. 0.5 mL of 4 mg/mL "magnesia mixture" is added, then 6.5 ml of 15% ammonia solution is stirred in slowly. This forms the solid magnesium ammonium arsenate, $MgNH_4AsO_4·6H_2O$. The mixture is left to precipitate for 30 min and is then centrifuged for 4000 rpm for 5 min. 025 ml of supernatant with the remaining free Mg^{2+} ions is diluted to 50 mL with 0.1 N HCl; this is the sample measured in the flame AA as above. The "magnesia mixture" is a magnesium ammonium solution prepared following the procedure by Vogel[21]. In short, 25 g of $MgCl_2$ and 50 g of NH_4Cl were dissolved in 250 mL of water. The product was left over night to precipitate, an the solid was removed by filtration. The solution was acidified by HCl and diluted to 500 mL with water.

Results and Discussion

In this project a resin for heavy metal ion remediation is being developed that is specific enough to take out all of the heavy metal ions, but still allows for the unspecific removal of other compounds the resins are generally used for. To achieve both goals in a cost-effective manner, inexpensive resins were chosen and then modified with a general method to increase their capacity. The amount of remval will be described in % removal and their dependence on the original concentration will be discussed. The quoted values are not actually total capacities; they could be described more accurately as "minimum capacities". We chose this method to quickly determine if these resins have high excess capacity and thus will be effective in the presence of competing ions even if the other ions are binding more strongly.

A flexible and simple method was needed that can be used for any compound and at the same time didn't increase cost greatly. Imprinting polymerization (Fig. 1) was chosen because one can imprint with any molecule, and it didn't require any additional synthesis. The only additional step for this method is a washing step that removes the initial templates.

Imprinting is a flexible method that can be used for all impurities[16]. It is even possible to imprint with mixtures of, for example, different heavy metal ions; it would be possible as well to use concentrated wastewater streams for imprinting (after removing all particles present). In that case the facility could remove all problematic compounds from their wastewater at the same time. Also, imprinted resins are reusable; the same initial washing step can be used to remove all impurities.

In this section the details for the imprinting polymerization for both the hydrophilic and the hydrophobic imprinting polymerizations will be discussed first, before the data for removal efficiency of several heavy metal ions is being presented.

Imprinting Polymerization for Hydrophilic Heavy Metal Ions

For hydrophilic heavy metal ions remediation a hydrophilic polymer was chosen for imprinting. Materials that are already used as coagulants in water remediation were imprinted; this makes it possible for the resin to act as specific and non-specific coagulant at the same time. Polyacrylamide was initially chosen and investigated and was imprinted with cadmium chloride[22].

It is actually more difficult to imprint with something as small as metal ions; in fact it was considered impossible. But this investigation showed quickly that it was possible; in fact some optimization yielded excellent results. What will not be possible, though, to make the imprinted resin completely specific for one ion, especially, since the imprinting polymerizations are being performed in water, which reduces the non-covalent bond strength. Therefore, resins with a large access of binding sites for heavy metal ions are being developed; this allows for the expected unspecific binding additional to the specific metal ion binding.

Polyacrylamide was found to be water soluble and therefore not completely suitable for these purposes[22]. Thus it was decided to use the slightly more hydrophobic polmethacrylamide. Also, it was considered advantageous to add negative charges to the molecule to additionally take advantage of the stronger ionic bonds. Therefore, first polymethacrylate (Fig. 3) was used, then polymethacrylamide was copolymerized with polyacrylamide (Fig. 3). The best results were obtained with a mixture of 70:30 methacylate:methacrylamide (Fig. 3)[22]. It was chosen to make it a random polymer structure, i.e. the methacrylamide and methacrylate repeating units are randomly distributed along the polymer chain. This is advantageous because the polymers chains will not order to form crystalline regions; crystalline polymers have lower surface area and thus less surface for the metal ions to bind.

Imprinting polymerization involves the preparation of a polymer in the presence of an template, in this case metal-ligand complexes, which is then removed[16]. A cavity specific for the metal-ligand complex remains (Fig. 1). In this case an unusual exchange resin is being imprinted; the ion-exchange resin is unusual in that it contains not only anionic groups but also a significant number of functional groups that can act as ligands to the metals, resulting in a different set of non-covalent bonds. Ionic and metal-ligand forces are some of the strongest non-covalent forces. This is important because in this case the imprinting polymerization was performed in water, which reduces the strength of all non-covalent bonds and thus reduces the specificity of the imprint[16]. Imprinted polymers for use in wastewater, however, should be made in water instead of made in organic solvent and then transferred; the imprint could swell along with the polymer, in sme cases making it no longer specific for the template. Although the specificity will not be as high, it will still be sufficient for remediation resins if there is excess binding capacity.

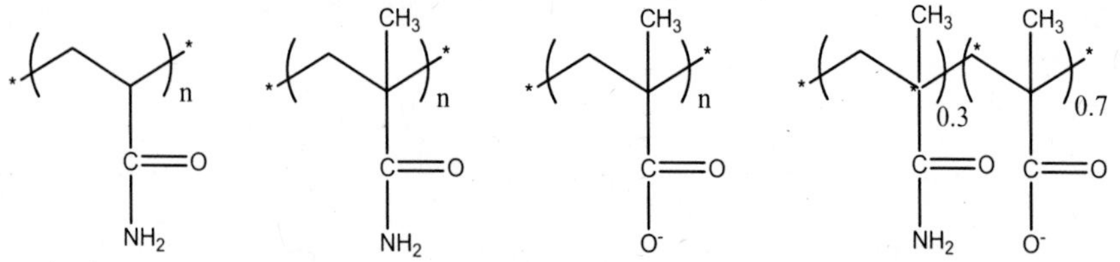

Polyacrylamide | Polymethacrylamide | Polymethacrylate | 70:30 Poly(methacylate-co-methacrylamide)

Figure 3. Hydrophilic Polymer Structures

Polymer chains are inherently flexible and need to be crosslinked in order to keep the structural integrity of the imprint after the removal of the template. At the same time, the crosslinking density has to be low enough to allow for complete removal of the template before use. It was found that 0.1 mol% of crosslinker in comparison to monomer was sufficient to crosslink the sample while allowing all templates to be removed before use. Higher concentrations resulted in template remaining in the resin; lower concentratons did not show much effectiveness in increasing the capacity of the resin.[22]

The resins were synthesized by radical polymerization under nitrogen atmosphere initiated by UV light, a common procedure for all polyacrylate-type resin. 2,2-Azobis(2-methylpropionamidine) dihydrochloride (AAPD), and ethylene glycol diacrylate (EDMA) were the water soluble radical initiator and crosslinker, respectively, the most commonly used initiator and crosslinker for hydrophilic resins. The templates were removed by dialysis, ultrafiltration could be used as well.

Again, it is important to create a large excess of metal binding sites to allow for the removal of all heavy metal ions in the presence of all competing molecules that will bind non-specifically. Therefore a resin with high surface area, and thus porosity, is needed. To increase the porosity of the resin, nitrogen gas was bubbled through the polymerization reactions.

For all resins that were tested, four samples were made. There were the imprinted resins as well as the controls: the synthesis was the same but the imprinted sample had the templates present during polymerizations, the controls did not. Then there were resins with increased porosity and resins with "regular" porosity: to make the former, nitrogen was bubbled through the reaction mixture during polymerization (called "porous samples"), to make the latter, nitrogen was introduced above the solution (called "non-porous samples"). With those four samples, the effect of both surface area and imprinting on the remediation efficiency was being investigated.

To prove that the polymerization occurs in the presence of any of the metal ions, IR spectroscopy was performed. In all cases the double bond from the monomers was not seen anymore in the polymer spectra, proving that radical polymerization is effective in the presence of heavy metal ions.

Since the crosslinked polymers were insoluble, their molecular weight cannot be determined by Gel Permeation Chromatography (GPC) or NMR end-group analysis. Differential scanning calorimetry (DSC) was used to compare glass transition temperatures (Tg). Because the Tg is dependent on the molecular weight and the molecular weight distribution in the low molecular weight range, this can be used as a rough comparison of sample molecular weights, even though crosslinking broadens the Tg transitions. All Tg were close to 40 °C ($\pm$ 1 degree), indicating that the molecular weight distributions of all samples was similar.

The transition in thermal gravimetric analysis (TGA) is dependent on the molecular weight and the molecular weight distribution as well. The spectra showed that the samples became more inhomogeneous upon crosslinking, but this homogeneity was repeatable upon imprinting and changes in porosity (Table 1). The results demnstrated that different batches, even with different templates, were consistent and repeatable. There was an inhomogeneity that was seen in all samples: about 10% of the resin was only lightly crosslinked, resulting in the small fraction that is thermally stable only to ca. 200 °C. These results also proved that all resins exceed the thermal stability required for water remediation applications.

Table 1. TGA data for imprinted 30:70 methacrylate:methacrylamide random copolymer and copolymer controls

TGA	CsCl in 70:30 poly (methacrylate-co methacrylamide)	CdCl₂ in 70:30 poly (methacrylate-co-methacrylamide)
Nonporous Control	221 °C (12%), 439 °C (88%)	202 °C (11%), 451 °C (89%)
Nonporous Imprinted	222 °C (9%), 414 °C (90%)	196°C (16%), 428 °C (84%)
Porous Control	229 °C (8%), 428 °C (90%)	201 °C (10%), 429 °C (88%)
Porous Imprinted	224 °C (5%), 416 °C (90%)	198 °C (8%), 428 °C (90%)

Imprinting Polymerization for Hydrophobic Heavy Metal Ions

For adsorption of hydrophobic contaminants a hydrophobic polymer is needed. Emulsion polymerization has been developed to synthesize hydrophobic polymers in water, a common method in industry (Fig. 2). Surfactant is added to the hydrophobic monomers in water, forming micelles. The surfactant molecules stabilize hydrophobic monomer droplets in water due to their internal, hydrophobic environment, and polymerization occurs within these droplets.

To our knowledge, imprinting emulsion polymerization had only been reported where a core-shell particle was being imprinted on the surface[18]. In this research, not only the surface but also inside pores were imprinted by metal ion complexes similarly to the imprinting of hydrophilic polymers described before.

Usually the surfactants used in emulsion polymerizations are ionic, since charge repulsion further increases the stabilization of the monomer droplets. In this case, though, non-ionic surfactants had to be used, otherwise the metal ions would have been retained on the surface of the droplets and no imprinting inside the monomer droplet would have occured. Therefore a non-ionic alkyl-oligo(ethylene glycol) surfactant was chosen for the polymerization.

Poly(methyl methacrylate) (PMMA, Fig. 4) was first imprinted during an emulsion polymerization, since PMMA is a common hydrophobic resin for wastewater treatment. Unfortunately, due to the high crystallinity of this polymer the surface area was very low (Table 2). Therefore, a random copolymer of 50:50 methyl methacrylate:phenyl methacrylate was synthesized. Again, two porous and two non-porous samples were prepared. Again, it was chosen to perform a a random copolymerization to reduce crystallinity and thus increase surface area.

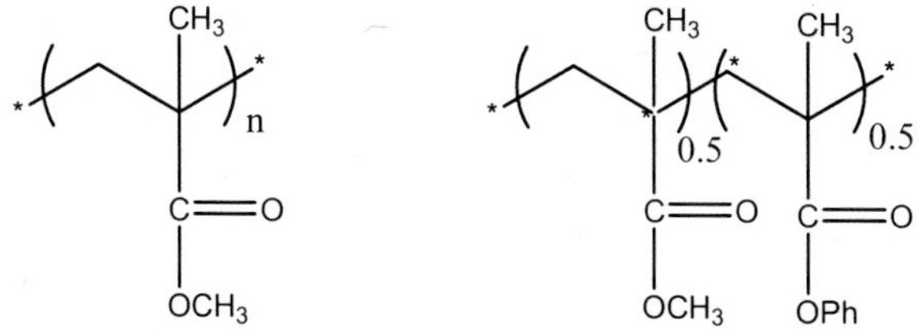

Poly(methyl methacrylate)   50:50 Poly(methyl methacylate-co
                                   -phenyl methacrylate)

Figure 4. Hydrophobic Polymer Structures.

The template for the hydrophobic case was the hydrophobic mercury dithizonate. This metal-ligand complex served two purposes: it is really hydrophobic, and it is a mercury complex that is safe to work with in a regular organic chemistry "wet laboratory". This complex also binds predominantly with weak van der Waals forces. Van der Waals forces are actually stronger in water, due to the so-called "hydrophobic effect", but still in general a lot weaker than the polar non-covalent forces.

Table 2. Surface Area of PMMA Samples.

PMMA	Non-porous Control	Non-porous Imprinted	Non-Porous Imprinted Washed	Porous Control	Porous Imprinted	Porous Imprinted Washed
BET Surface Area (m^2/g)	2.9377	0.7946	4.55	3.1683	0.1333	0.15

Heavy Metal Remediation

The first template chosen in this study was cadmium chloride, due to its high toxicity, frequent occurrence in wastewater[23], and its common occurrence in the Great Lakes. It is emitted from industries such as mining and steel manufacturing[23].

In both the porous and non-porous samples, the imprinted polymer (which still contained cadmium) proved denser and less porous than the respective non-imprinted polymer. Since porosity changes the surface area of the sample and thus possible metal uptake, it was important to measure the porosity of each sample. The highest surface area was found in the porous imprinted sample, even though in this sample still retained $CdCl_2$. Overall the porosity is less then the suggested 10 m^2/g, but the resins show a large capacity nonetheless (see below), thus this surface area might be sufficient.

Table 3. Porosity of CdCl$_2$ Resin Samples

Polymer sample	Non-Porous Control	Non-Porous Imprinted	Porous Control	Porous Imprinted
BET Surface Area (m^2/g)	0.27	0.30	0.25	3.95

Cadmium adsorption was measured by using the different polymer samples as the solid phase of a chromatography column. Known concentrations of cadmium chloride in water were passed through these columns and the amounts of cadmium in the elute and in the solid phase were measured by atomic absorption. In all cases, the imprinted polymers were compared with the non-imprinted controls to determine the effect of imprinting on cadmium adsorption. Since all of the studied resins were negatively charged at neutral pH, all retained cadmium due to electrostatic forces (i.e. all resins acted as ion exchange resins). In this project it was determined if imprinting increases adsorption due to the creation of additional specific binding pockets for cadmium ions based on metal-ligand forces. The concentrations used to determine the effectiveness of the resin were in large excess of the concentrations that are usually found in water: 20, 40, and 80 ppm of CdCl$_2$. The excess binding sites will be covered by random adhesion of other impurities in the water.

In both porous and non-porous samples the imprinted polymer demonstrated increased adsorption in comparison to the non-imprinted controls, in fact only 200 mg of imprinted sample removed 100% of 80 ppm of CdCl$_2$ (Table 4)! The non-porous, imprinted polymer showed complete adsorption for 20 ppm, 40 ppm, and 80 ppm CdCl$_2$ with only 50 mg of sample, thus they have at least a four-fold capacity compairedd with the non-porous imprinted samples. Unfortunately, these samples experienced more swelling than the porous samples, to such a degree that the CdCl$_2$ solution had to be forced through the chromatography columns with pressure. The porous, imprinted samples showed clear improvement over the control samples but were not as effective as the non-porous, imprinted samples. On the other hand, this polymer was not tested at equally high concentrations. Therefore, it is possible that equally high adsorption can be reached with the non-porous as with the porous samples.

Table 4. CdCl$_2$ remaining after column

ppm CdCl$_2$	% CdCl$_2$ remaining after column, 200 mg resin (non-porous imprinted: 50 mg resin)			
	Non-Porous Control	Non-Porous Imprinted	Porous Control	Porous Imprinted
20	0	2.3% (0.5ppm)	1.0% (0.2ppm)	0
40	1.0% (0.4ppm)	0	1.0% (0.2ppm)	0
80	18.0% (14.4ppm)	0	26% (20.4ppm)	0

The same 70:30 methacrylate:methacrylamide random copolymer was also imprinted with cesium chloride. Again, the two non-porous and porous samples were prepared to study the effect of porosity on metal adsorption. In all cases, even though the capacity for cesium ions was high, no significant difference in metal adsorption between the imprinted polymers and the non-imprinted controls was found up to 180 ppm of cesium chloride (Table 5). Only up to 95% CsCl were removed at the low concentrations, at the high concentrations up to 50%. The adsorption of cesium chloride ions was also measured in the presence of sodium ions. Again, no significant difference in metal adsorption between the imprinted polymers and the non-imprinted controls was found, and the presence of sodium ions reduced the adsorption of cesium ions significantly. Cesium ions, contrary to cadmium ions, bind mostly via ionic forces, not metal-ligand forces. Therefore, this result shows that imprinting with heavy metal ions only increases adsorption if metal-ligand forces are used for binding. This is understandable, since ionic forces in water are greatly reduced.

Table 5. CsCl$_2$ remaining after column

ppm $CsCl_2$	% CsCl$_2$ remaining after column, 200 mg resin			
	Non-Porous Control	Non-Porous Imprinted	Porous Control	Porous Imprinted
20	7.0% (1.4ppm)	12% (2.5ppm)	7.0% (1.4ppm)	5.0% (1.0ppm)
40	9.0% (3.6ppm)	3.5% (1.4ppm)	5.0% (2.0ppm)	3.5% (1.4ppm)
80	13% (10ppm)	16% (13ppm)	2.0% (1.6ppm)	2.5% (2.0ppm)
180	47% (85ppm)	46% (83ppm)	53% (95ppm)	49% (89ppm)

Initial data for arsenic trioxide was collected as well (Table 6). With arsenic, both the As^{3+} and As^{5+} ions are toxic. We will eventually imprint with both ions and do competitive studies with both ions to determine if the imprinted resins can distinguish between both ions. Here initial data for As^{3+} is presented. The porous, imprinted polymer was for some reason denser than the other samples, which likely explains the low efficiency of that specific resin. We are currently repeating those results.

Mercury (II) dithizonate was used to test the new imprinting emulsion polymerization method for hydrophobic compounds. Unfortunately the porosity of these polymers was still very low, even though a random copolymerization of a mixture of monomers, methyl methacrylate and phenyl methacrylate, was used. This is likely because both polymethacrylate and polyphenyl methacrylate homopolymers are highly crystalline.

Table 6. Arsenic trioxide remaining after column

AsO_3 ppm added	Arsenic Trioxide Remaining in Solution ppm			
	Non-Porous Control	Non-Porous Imprinted	Porous Control	Porous Imprinted
20ppm	1.70 (8.5%)	0.74 (3.7%)	1.35 (5.8%)	6.72 (33.6%)
40ppm	2.07 (5.2%)	1.25 (3.1%)	3.95 (9.9%)	10.9 (27.3%)
80ppm	6.90 (8.6%)	3.44 (4.3%)	4.42 (5.5%)	15.7 (19.6%)

Adsorption data of mercury (II) dithizonate was measured in spite of the low surface area, and the data clearly shows that imprinting and porosity increases adsorption (Table 7). Again, 200 mg of the porous, imprinted resin can retain 80 ppm of mercury dithizonate, a 100% removal rate.

68

Table 7. Mercury dithizonate remaining after column

ppm Hg^{2+}	% Mercury dithizonate remaining after column, 200 mg resin			
	Non-Porous Control	Non-Porous Imprinted	Porous Control	Porous Imprinted
20	31 % (6.2ppm)	0	38 % (7.6ppm)	0
40	18 % (7.2ppm)	0	17 % (6.8ppm)	0
80	10 % (8.0ppm)	5 % (4.0ppm)	8.0 % (6.4ppm)	0

Summary and Future Work

The data presented here demonstrate that imprinting polymerization is a simple, yet effective and general method to significantly increase the binding capacity of common resins for heavy metal ions. It was found that increased surface area results in increased adsorption of heavy metal ions. Imprinting polymerization did not increase metal ion adsorption when metal ions were used that primarily bind via ionic forces, not metal-ligand forces.

In this paper an easy method for imprinting emulsion polymerization inside the monomer droplet was reported as well. The crucial difference in polymerization procedure was found to be the use of a non-ionic surfactant during the polymerization. With this procedure, imprinting polymerization can now be used for remediation of any hydrophilic or hydrophobic compound in water. In fact, it could also be used to make a resin specific for several contaminants in a specific wastewater stream by just using a concentrated version of that stream (with all particles removed) for the imprinting polymerization.

In the future, more work will be done on both toxic arsenic ions, as well as on further increasing the surface area of the hydrophobic polymer. Also, total capacity and adhesion isotherms will be determined for all samples.

Acknowledgements

The authors thank Katelyn Carter for her technical assistance with the DSC and TGA samples. The funding by the Central Michigan University President's Research Investment Fund and Clarkson University is gratefully acknowledged.

References

1. Gregory J., The Action of Polymeric Flocculants. In *Flocculation, Sedimentation, and Consolidation*, Moudgil B.M.; Somasundaran P., Eds. Engineering Foundation: The Cloister, Sea Island, Georgia, 1985; pp 125-137.
2. Koltuniewicz Andrzej B.; Drioli Enrico, *Membranes in Clean Technologies*. Wiley-VCH: Weinheim, 2008.
3. Manju G.N.; Krishnan K.A.; Vinod V.P.; Anirudhan T.S., An Investigation into the Sorption of Heavy Metals from Wastewaters by Polyacrylamide-

Grafted Iron(III) Oxide. *Journal of Hazardous Materials* **2002**, 91, (1-3), 221-238.

4. Xiang L.; Chan L.C.; Wong J.W.C., Removal of Heavy Metals from Anaerobically Digested Sewage Sludge by Isolated Indigenous Iron-Oxidizing Bacteria. *Chemosphere* **2000**, 41, 283-287.

5. U.S. Environmental Protection Agency Toxic Release Inventory (TRI). http://www.epa.gov/tri/

6. U.S. Environmental Protection Agency Consumer Factsheet on: CADMIUM. http://www.epa.gov/safewater/contaminants/dw_contamfs/cadmium.html

7. Wright Karen, Our Preferred Poison. *Discover* **2005**, (3), 58-65.

8. Wikipedia Arsenic Poisoning. http://en.wikipedia.org/wiki/Arsenic_poisoning (November 18),

9. Green Facts Scientific Board; International Programme of Chemical Safety Scientific Facts on Arsenic. http://www.greenfacts.org/en/arsenic/1-2/arsenic-2.htm

10. Shubha K.P.; Raji C.; Anirudhan T.S., Immobilization of Heavy Metals from Aqueous Solutions Using Polyacrylamide Grafted Hydrous Tin(IV) Oxide Gel having Carboxylate Functional Groups. *Water Research* **2001**, 35, (1), 300-310.

11. Li D.; Zhu S.; Pelton R.H.; Spafford M., Flocculation of Dilute Titanium Dioxide Suspensions by Graft Cationic Polyelectrolytes. *Colloid and Polymer Science* **1999**, 277, 108-114.

12. Wang C.-C.; Chen C.-Y.; Chang C.-Y., Synthesis of Chelating Resins with Iminodiacetic Acid and its Wastewater Treatment Application. *Journal of Applied Polymer Science* **2002**, 84, 1353-1362.

13. Cimino G.; Passerini A.; Toscana G., Removal of Toxic Cations and Cr(VI) from Aqueous Solution by Hazelnut Shell. *Wat. Res.* **2000**, 34, (11), 2955-2962.

14. Zuman P.; Rupp E., Lignin as Adsorbent and Detoxicant. *International Journal of Environmentally Conscious Design and Manufacturing* **2001**, 10, (1), 23-30.

15. Cimino, G.; Caristi, C., Acute Toxicity of Heavy Metal Ions to Aerobic Digestion of Waste Cheese Whay. *Biological Wastes* **1990**, 33, 201-210.

16. Yan M.; Ramstroem O., *Molecularly Imprinted Materials, Science and Technology*. Marcel Dekker: New York, 2005.

17. Blackley D.C., *Emulsion Polymerization: Theory and Practice*. Applied Science Publishers: London, 1975.

18. Perez N.; Whitcombe M.J.; Vulfson E.N., Molecularly Imprinted Nanoparticles Prepared by Core-Shell Emulsion Polymerization. *J. Appl. Polym. Sci.* **2000**, 77, 1851-1859.

19. Mercado C.; Mosey J.; Ashraf S.A.; Mueller A., Imprinted Polymers for the Removal of Hydrophobic Compounds from Wastewater. *Environmental Preprints* **2008**, 48, (1), 489.

20. Hassan S.S.M.; Eldesouky M.H., Indirect Microdetermination of Arsenic Compounds by Atomic Absorption Measurements. *Zeitschrift der Analytischen Chemie* **1972**, 259, 346-348.

21. Vogel Arthur I., *Quantitative Inorganic Analysis*. Longmans: London, 1961; p 501.
22. Eastman C.; Goodrich S.; Gartner I.; Mueller A., Imprinted Polymers in Wastewater Treatment. *Environmental Preprints* **2004,** 45, (2), 404-405.
23. EPA Toxic Release Inventory (TRI). http://oaspub.epa.gov/enviro/fii_master.fii_retrieve?city_name=Detroit&state_code=MI&all_programs=YES&program_search=1&report=1&page_no=1&output_sql_switch=TRUE&database_type=TRIS

Chapter 5

Gamma Radiation-Polymerized Metal Methacrylates for Adsorption of Metal Ions from Wastewater

Bryan Bilyeu[1], Carlos Barrera-Díaz[2], and Fernando Ureña-Nuñez[3]

[1]Department of Chemistry, Xavier University of Louisiana, 1 Drexel Drive, New Orleans, LA 70125
[2]Facultad de Química, Universidad Autónoma del Estado de México, Toluca, Estado de México, México
[3]Instituto Nacional de Investigaciones Nucleares, México, D.F., México

The properties of the functional groups on a polymer are what determine how well it can adsorb metal cations and/or oxyanions from a solution. Oxygen-containing functional groups are effective adsorption sites, but modifications and environment affect their strength, so we synthesized polymethacrylates (PMA) with metals in the chains adjacent to the carboxyl groups. In separate studies we have evaluated Zn(II)PMA(1) and Cu(II)PMA(2) for adsorption of Pb^{2+} and Fe(II)PMA and Fe(III)PMA(3) for adsorption of hexavalent chromium in the form of CrO_4^{2-}. The metal methacrylate monomers were polymerized with gamma radiation and characterized for structure and properties. These synthetic sorbents were used in batch sorption tests to determine the kinetics and mechanism of the processes, as well as their capacities. In the lead sorption studies, Zn(II)PMA exhibited heterogeneous Freundlich behavior adsorbing 94% of the lead in a 25 ppm solution and 67% in a 150 ppm solution with a capacity of 20.11 mg per gram of sorbent, while Cu(II)PMA matched the homogeneous Langmuir model adsorbing 56% from a 75 ppm solution and 48% from a 150 ppm solution with a capacity of 6.22 mg per gram of sorbent. The Zn(II)PMA has a higher capacity for lead cations than many adsorbents used in wastewater treatment. In the Cr(VI) oxyanion studies, Fe(II)PMA exhibited a heterogeneous

Freundlich behavior adsorbing 70% of the chromate ions in a 75 ppm solution with a capacity of 22.26 mg per gram of sorbent, while Fe(III)PMA followed a homogeneous Langmuir mechanism adsorbing only 30% of a 75 ppm solution with a capacity of 3.52 mg per gram. The Fe(II)PMA has a high capacity for chromate ions and shows great promise as an effective sorbent.

Introduction

Most heavy metal ions are toxic or carcinogenic and hence present a threat to human health and the environment when they exist in or are discharged into various water resources. Heavy metal pollution exists in wastewater discharge of many industries among which the plating facilities, tanneries and mining operations are easily distinguishable due to their severe environmental impacts and ever present risks associated with mismanagement. While most metal ions in solution are found as cations, some like chromium and arsenic also exist as oxyanions, which affect the adsorption mechanisms(4).

Adsorption is one of the methods commonly used to remove heavy metal ions from aqueous solutions(5). The efficiency of adsorption relies on the capability of the sorbent to concentrate metal ions from the solution onto its surface. There are many types of adsorbents, including both inorganics like activated carbon(6), metal oxides(7), and minerals(8,9), and organic polymers like cellulose(10,11) and chitin(12) biosorbents(13). One of the most promising methods for heavy metal removal is the adsorption of pollutant ions onto natural and synthetic polymeric materials, which usually are abundant and inexpensive(14). Many natural cellulosic materials like *Opuntia* (Prickly Pear Cactus)(15), peat(16), seaweed(17), algae(18), plant roots(19), carrots(20) and many others are effective sorbents. Moreover, after these inexpensive sorbents have been expended, they can be easily disposed or regenerated. Modifications to the functional groups on polymers can enhance effectiveness(21), as we have shown with *Opuntia*(22,23) and others have shown with sawdust(24), coconut coir pith(25), as well as synthetic polymers like polyacrylonitrile(26). Due to the relatively large external specific surface areas, fibers are the preferred form for adsorbents. In the adsorption process, metal ions in the aqueous solutions may be transported through diffusion or convection to the surface of the adsorbent and then become attached to the surfaces due to various physical or chemical interactions between the metal ions and the surface functional groups of the adsorbent.

Polymer production using irradiation techniques presents the following advantages over traditional methods: the synthesis is carried out in the absence of catalysts and initiators and polymerization and crosslinking may occur simultaneously(27). Furthermore, there is no need to add solvents to perform the polymerization. Thus, this technique could be considered as a clean way to obtain polymeric materials. While various wavelengths are used in chemical reactions, gamma radiation is well suited for polymerizations since it interacts with alkene bonds. The obvious disadvantage is the need for a gamma irradiation facility. However, the recent interest in gamma irradiation of foods to prevent bacterial contamination and in nuclear energy to reduce global warming is expected to increase the availability of facilities and source material.

In evaluating adsorbents for effectiveness, the material itself must be fully characterized before the detailed adsorption studies of rate and capacity. Our work in gamma-polymerized metal methacrylates has included zinc(II), copper(II), iron(II) and iron(III) polymethacrylates to remove metal cations (lead) or oxyanions (hexavalent chromium) from industrial wastewater. The metal polymethacrylates were used as sorbents in a series of batch experiments to investigate its capacity in removing the metal ions from aqueous solutions. In order to characterize the material composition and the mechanisms involved, the following techniques were used: scanning electron microscopy (SEM), energy dispersion analysis (EDX), electron paramagnetic resonance (EPR), X-ray diffraction, Fourier transform infrared spectroscopy (FT-IR) and X-ray photoelectron spectroscopy (XPS).

Materials and methods

Synthesis of Metal Methacrylate Monomers

Metal methacrylates were synthesized in the following steps: an aqueous solution of $NaHCO_3$ was treated with methacrylic acid and the mixture was stirred for 30 minutes (Eq. (1)), then the metal chloride (MCl_2 or MCl_3) was added and stirred again for one hour at 40 °C (Eqs. (2,3)). Once the reaction took place, the insoluble metal methacrylate precipitate was filtered out, washed with distilled water and dried under vacuum.

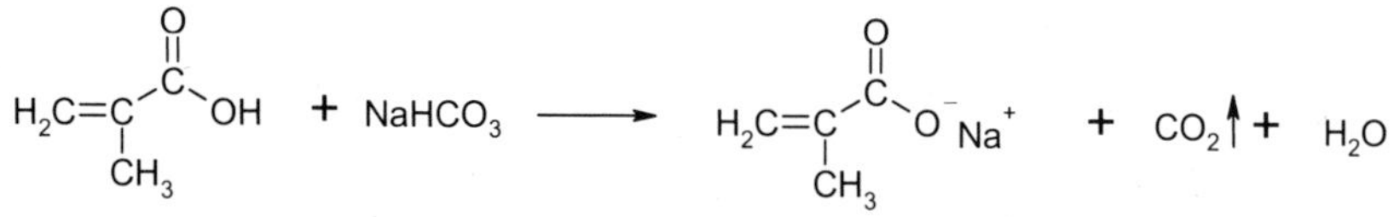

Eq. (1)

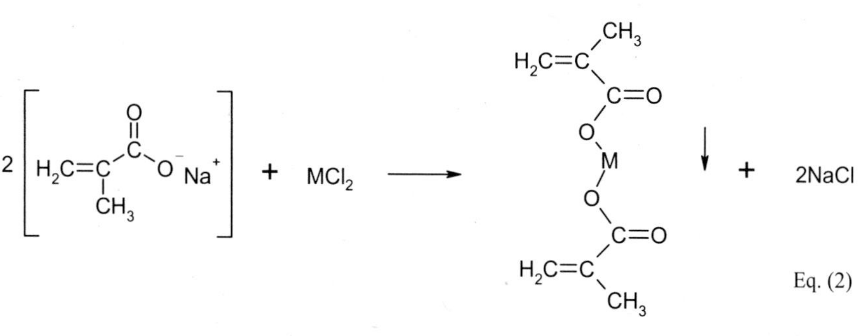

Eq. (2)

$$3 \left[H_2C = C \begin{matrix} \\ CH_3 \end{matrix} \overset{O}{\underset{}{C}} - O^- \, Na^+ \right] + MCl_3 \longrightarrow \quad + 3NaCl \qquad \text{Eq. (3)}$$

Polymerization of Metal Methacrylates

The γ-ray induced polymerization of the monomer was carried out in a gamma irradiation unit ALC gammacell-220, supplied with a ^{60}Co source. A 20 kGy dose was applied at a 0.5 kGy h^{-1} rate. It has been shown elsewhere by our group(28), that these conditions induce complete polymerization of the monomer with the greatest crystallinity index (CI). The polymerization reactions for 2 and 3 coordinate metals are shown in Equations 4 and 5.

Eq. (4)

Eq. (5)

Monomer and Polymer Characterization

Scanning Electron Microscopy (SEM) and Energy Dispersive X-ray Analysis (EDX)

The SEM characterization was carried out on samples of both monomer and polymer, using a JEOL JSM-5900 LV microscope to obtain information on the composition and general features of the structures. Scanning electron microscopy provides secondary electron images of the surface with resolution in the micrometer range, while energy dispersive X-ray spectroscopy offers *in situ* chemical analysis of the bulk. The chemical composition of the polymer was determined by a DX-4 analyser coupled to the SEM, before and after contact with the aqueous solution.

Electron paramagnetic resonance (EPR)

The polymers were analyzed using electron paramagnetic resonance (EPR) to confirm the presence of free radicals during and after the polymerization. This study was done with a Varian E-15 spectrometer operating at the X-band of the microwaves and was recorded as the first derivative of absorption spectrum. All measurements were performed at room temperature and the instrument settings were as follows: magnetic field 330 mT, scan range 40 mT, scan time 8 minutes, magnetic field modulation amplitude 0.1 mT, modulation frequency 100 kHz, microwave power 2.0 mW (nominally 1.0 mW per half of the dual cavity); receiver gain and time constant were adjusted according to the signal intensity.

X-ray Diffraction

The crystallinity of the metal polymethacrylates were analyzed with an X-ray diffractometer scanning in the 2θ range 0-60. Copper radiation was used with a diffracted beam monochromator tuned to $K\alpha$ radiation

Fourier Transform Infrared Spectroscopy (FTIR)

The monomers and polymers were analyzed with a Nicolet Magna-IR 550 to observe the changes in the chemical bonds and structure and to assure that polymerization had taken place.

Thermogravimetric analysis (TGA)

Analysis of the thermal stabilities were performed on a TA Instruments TGA 51 thermogravimetric analyzer, which was operated in a nitrogen atmosphere at a heating rate of 10 °C min^{-1} from 25 to 800 °C.

X-ray Photoelectron Spectroscopy (XPS)

XPS analyses of the metal methacrylates before and after the lead or chromium adsorption was carried out on an AXIS HIS spectrometer (Kratos Analytical Ltd., U.K.) with an Al kα X-ray source to determine the atoms present on the surface of the polymers.

Surface area measurements

The polymer surface areas were determined by standard multipoint techniques using a Micromeritics Gemini 2360 instrument. Prior to analysis the samples were dehydrated at 80 °C for 2 hours.

Adsorbtion of Pb(II) or Cr(VI) on Metal Polymethacrylates

In order to evaluate the metal ion removal capacity of the polymers, batch equilibrium tests were conducted at constant temperature (18 ± 0.5 °C). The powdered metal polymethacrylate samples were put in contact with the aqueous Pb(II) or Cr(VI) solutions. All solutions were prepared with analytical grade reagents, using deionized water (18 MΩcm resistivity). The mixtures were stirred, then the phases were separated by filtration and the Pb(II) or Cr(VI) in solution was evaluated. The selected parameters: mass/volume ratio, initial metal concentration and contact time were studied. Duplicate experiments permitted averaging of results.

Quantification of metal ion concentration in solution.

The concentration of metal ions in solution, before and after the sorption process was determined using a Perkin Elmer 2380 Atomic Absorption Spectrophotometer. All calibrations and procedures were carried out in accordance with AWWA standards(29).

Ionic Species Distribution

Pb(II) and Cr(VI) form different complexes in aqueous solution depending on the pH, so distribution diagrams of the chemical species present were calculated using the MEDUSA program.

Results and discussion

Materials Characterization

SEM and EDX analysis of Zn(II)MA and Zn(II)PMA

The Zn(II)MA monomer precipitates as thin laminar fibers (Figure 1), while Cu(II)MA forms puffy clumps of short fibers (Figure 2), Fe(III)MA forms small spherical granules (Figure 3), and Fe(III)PMA forms a network of platelets (not shown). After gamma polymerization of the solid monomers, the structures tend to split and splinter, as shown for Zn(II)PMA in Figure 4.

While imaging the monomers and polymers in the SEM, Energy Dispersive X-ray elemental analysis was done to confirm the elemental distribution expected from the reaction and to use for comparison with the polymer after sorption. The elemental composition of the Zn(II)PMA fibers are compared to that expected from the chemical reaction in Table I. The slight change in composition due to the polymerization is shown for Cu(II)MA and Cu(II)PMA in Table II. The expected difference in composition between Fe(II)MA and Fe(III)MA is shown in Table III.

Figure 1. Secondary electron image of the Zn(II)MA monomer showing fibrous structure. The magnification marker is 50 μm. (Reproduced from reference 1. Copyright 2007 American Chemical Society)

Figure 2. Secondary electron image of the Cu(II)MA monomer showing puffy (short fiber) structure. The magnification marker is 10 μm.

Figure 3. Secondary electron image of the Fe(II)MA monomer showing granular structure. The magnification marker is 50 μm.

Figure 4. Secondary electron image of the Zn(II)PMA polymer showing the splintering of the fibers after irradiation. The magnification marker is 20 μm.

Table I Elemental Composition of Zinc Polymethacrylate

Compound	Elemental composition (Atomic %)		
	C	O	Zn
Zn(II)PMA	41.04	27.91	31.05
Theoretical	40.85	27.23	27.66

NOTE: Remaining composition is hydrogen which is not detectable by EDX.

Table II Elemental composition of Cu(II)MA and Cu(II)PMA

Compound	Elemental composition (Atomic %)		
	C	O	Cu
Cu(II)MA	55.04	31.07	12.89
Cy(II)PMA	51.48	34.37	14.15

NOTE: Remaining composition is hydrogen which is not detectable by EDX.

Table III Elemental composition of Fe(II)MA and Fe(III)MA

Compound	Elemental composition (Atomic %)		
	C	O	Fe
Fe(II)MA	40.99	28.80	30.20
Fe(III)MA	37.82	30.19	31.26

NOTE: Remaining composition is hydrogen which is not detectable by EDX.

Electron paramagnetic resonance (EPR)

The EPR spectra of the zinc polymethacrylate obtained at 20 kGy dose of radiation is shown in Figure 5. The signal intensities are complex, however, the peak at 328 mT indicates that the propagative free radical is of the type:

$$R - CH_2 - \dot{C} - (CH_3) \, COOH \qquad (6)$$
$$\quad\quad\;\; \beta \quad\;\; \alpha$$

In Zn(II)PMA, Cu(II)PMA, Fe(II)PMA and Fe(III)PMA, there are no indications of remaining free radicals after the polymerization is complete.

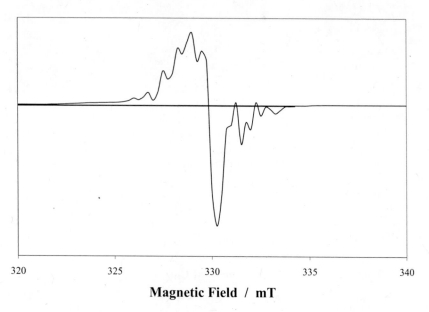

Magnetic Field / mT

Figure 5. EPR spectra of Zn(II)PMA: Signal at 330 indicates the presence of free radicals. (Reproduced from reference 1. Copyright 2007 American Chemical Society)

X-ray Diffraction

X-ray diffraction was used to determine the degree of crystallinity in the monomers and polymers. The X-ray spectra of Zn(II)MA and Zn(II)PMA are shown in Figure 6. The peak signals are well defined, with the largest peak at 7.5 degrees 2θ, followed in a lesser extent by signals at 15 and 22.5 degrees indicating a very crystalline structure. The spectra for Cu(II)MA and Cu(II)PMA were very similar.

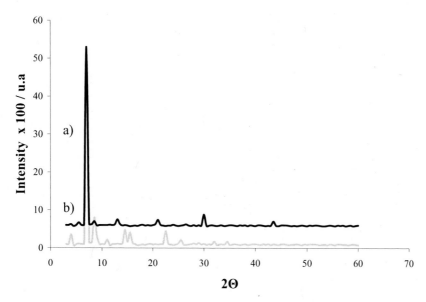

Figure 6. X-ray diffractogram of a) Zn(II)MA and b) Zn(II)PMA. Note that in both cases the peaks are clearly shaped indicating a crystalline array. (Reproduced from reference 1. Copyright 2007 American Chemical Society)

FT-IR analysis

Fourier Transform Infrared (FTIR) spectroscopy is used to identify the chemical bonds present. FTIR spectra of Zn(II)MA and Zn(II)PMA before and after the contact with a lead solution are shown in Figure 7. This technique was used to identify important functional groups. Figure 5a shows that the FTIR spectra of the monomer displayed a small band at 3080 cm^{-1} indicating alkene stretching, with the peak at 1860 representing the characteristic overtone of the double bond, while at 1640 there is confirmation of the carbon-carbon double bond. The five characteristics bands of a carboxylic acid are replaced by two bands in 1560 and 1430 cm^{-1}, which correspond to the conversion of the inorganic salt. At 2970 and 2930 cm^{-1} are signals corresponding to the symmetric and asymmetric movements of the C-CH$_3$ bond. Finally, the methyl signal at 1375 cm^{-1} is observed. On the other hand, it is observed in Figure 7b that there is no signal at 3080 cm^{-1}, indicating the polymerization has occurred. The FTIR analysis of Cu(II)MA and Cu(II)PMA showed similar bands, with the same implications.

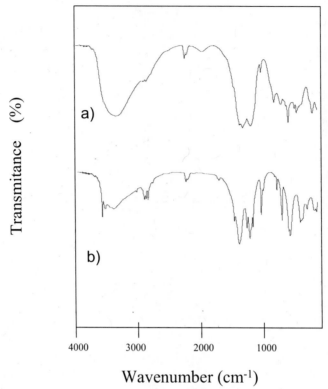

Wavenumber (cm⁻¹)

Figure 7. FTIR spectra of a) Zn(II)MA and b) the Zn(II)PMA. (Reproduced from reference 1. Copyright 2007 American Chemical Society)

Thermogravimetric Analysis (TGA)

Thermogravimetric analysis (TGA) indicates thermal stability and thermal decomposition onset temperature, as well as the weight loss due to decomposition. The weight % loss thermogram of Zn(II)MA and Zn(II)PMA shown in Figure 8 indicates that degradation of the monomer and polymer begin at 212 °C and 387 °C respectively. The monomer starts to degrade at a lower temperature than the polymer due to the increased thermal stability of crosslinking. However, after 500 °C both materials have a similar degradation. The analysis also indicates that the polymer can be consolidated (decomposed) to less than half its original weight for disposal. Cu(II)PMA shows similar behavior with degradation beginning around 208 °C and reaching less than half of the original weight by 600 °C.

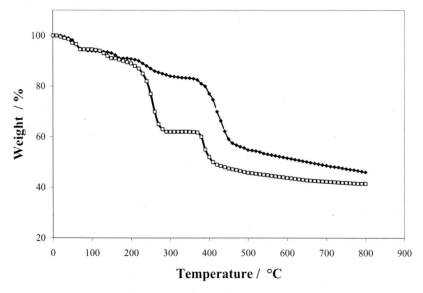

Figure 8. Weight loss on heating in a nitrogen atmosphere of the Zn(II)MA (□) and Zn(II)PMA (♦).The first weight loss corresponds to water loss from the material, while the second one indicates the onset of material degradation. (Reproduced from reference 1. Copyright 2007 American Chemical Society)

Surface area measurements

The result of the BET analysis of the surface area of the Zn(II)PMA fibers was 1.65 m² g⁻¹. This value is relatively small compared with carbon. However, sorption results indicate that good adsorption occurs onto this material. As expected from the shapes, the surface area of the Fe(II)PMA small spheres was an even higher 3.46 m² g⁻¹, while the large Fe(III)PMA structure had a much smaller surface area of 0.14 m² g⁻¹.

pH effect on ionic species in solution

Concentration and pH define the different ionic species present in aqueous solution. In Figure 9, the distribution of the chemical species in a 150 mg L⁻¹ lead aqueous solution as a function of pH is presented. Note that, there are two species, namely, Pb^{2+} and $Pb(OH)_2$. Lead will be presents as a free ion up to pH of 6, when the fraction of lead hydroxide present in aqueous solution has the equal fraction amount. The most important information that this diagram provide is that indicates that precipitation of lead will occur when the aqueous solution is above a pH of 6. Therefore, all lead sorption experiments were carried at a pH of 5.5.

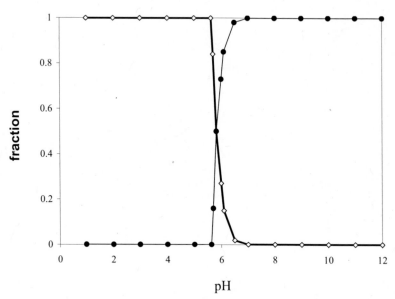

Figure 9. Predominant lead species in aqueous solution. [Pb] = 150 mg L⁻¹.
Pb²⁺ (◊) and Pb(OH)₂ (●).(Reproduced from reference 1. Copyright 2007
American Chemical Society)

The effect of solution pH values on the adsorption of lead ion on the polymer is shown in Figure 10. It can be observed that increasing the pH of the aqueous solution the lead absorption is increased. At a pH below 2 the lead adsorption is not detected; however the adsorption amount of lead ions onto polymer increased consistently for pH from 3 to 6.

Since hexavalent chromium is an oxyanion, the species present at different pH values are different than those for metal cations, as shown in Figure 11. Unlike typical metal cations, including Cr(III), the hexavalent chromium oxyanion is soluble at all pH values. However, the adsorption process itself does limit the pH range. At extremely low pH, the uncharged chromic acid species would have extremely weak adsorption. The functional groups on the polymer responsible for adsorption must be protonated to adsorb anions, so acidic conditions are necessary. In balancing these two effects, the optimum pH for adsorption of Cr(VI) oxyanions is around pH 6.

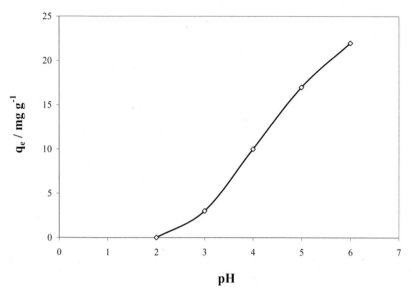

Figure 10. pH effect on lead adsorption. (Reproduced from reference 1. Copyright 2007 American Chemical Society)

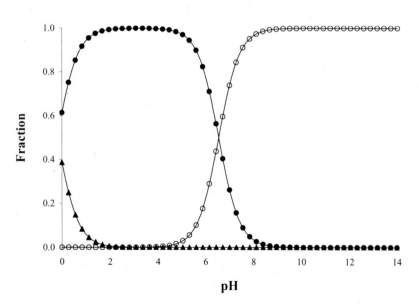

Figure 11. Distribution of species in a 5 mg L^{-1} solution of Cr(VI): ● $HCrO_4^-$, O CrO_4^{2-}, ▲ H_2CrO_4.

Adsorption Results

For the Zn(II)PMA adsorption of Pb(II), two initial sorbent amounts of 50 and 100 mg and two initial solution concentrations of 25 and 150 mg L^{-1} were evaluated at pH 5.5. The experimental plots for 100 mg of Zn(II)PMA sorbent in Pb(II) removal as a function of contact time for both initial concentrations are shown in Figure 12. Note that at a concentration of 150 mg L^{-1} Pb(II), the Zn(II)PMA achieved a maximum removal of 67% at 70 minutes, with no significant improvement afterwards. Similar behavior is observed at a concentration of 25 mg L^{-1}, but with a maximum 94% removal.

Cu(II)PMA was done similarly with 50 and 100 mg of sorbent and 75 and 150 mg L^{-1} of Pb(II) at pH 5.5. The results look similar to those of the Zn(II)PMA, except that the results for 100 mg of sorbent indicate a 56% removal for 150 mg L^{-1} was only 48% and for 75 mg L^{-1} was 56%.

Fe(II)PMA and Fe(III)PMA sorption of Cr(VI) was done with 100 mg of sorbent and a hexavalent chromium concentration of 75 mg L^{-1}. The Fe(II)PMA Cr(VI) removal graph looked much like the previous two, but achieved a maximum equilibrium removal of 70% at 60 minutes. The Fe(III)PMA only achieved a maximum equilibrium removal of 30%.

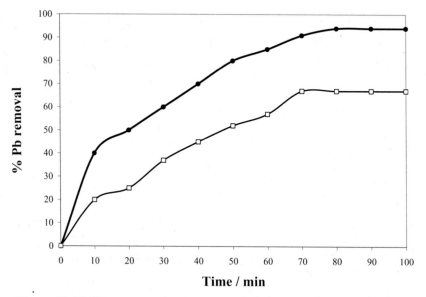

Figure 12. Pb(II) concentration in aqueous solution as a function of contact time for initial concentrations of 150 mg L^{-1} (□) 25 mg L^{-1} (●). (Reproduced from reference 1. Copyright 2007 American Chemical Society)

Adsorption Isotherms

Adsorption isotherms for all the sorbents were fitted to Langmuir and Freundlich equations in order to calculate the maximum adsorption capacity of the polymers. The Langmuir equation is based on the assumption of a structurally homogeneous adsorbent where all sorption sites are identical and energetically equivalent. It is assumed that once a metal ion occupies a site, no further adsorption takes place in this site. Langmuir constants q_0 (sorption capacity of the material, mg g^{-1}) and b (energy of adsorption) can be graphically obtained by plotting C_e/q_0 vs C_e, which has a slope of $1/q_0$ and a intercept of $1/q_0b$. C_e is the equilibrium concentration of the metal ion. The linear equation is shown in Equation 7.

$$C_e/q = (1/q_0)b + (1/q_0)C_e \qquad (7)$$

The Freundlich model assumes that the adsorbent consist of a heterogeneous surface composed of different adsorption sites. Freundlich parameters K_f (related to sorption capacity) and $1/n$ (intensity of the adsorption) can be obtained from the linearized plots of log q_e versus log C_e. Equation 8 shows the Freundlich isotherm model.[12]

$$\text{Log } q_e = \log K_f + 1/n \log C_e \qquad (8)$$

To determine the fit to the two models a series of batch experiments were done for each sorbent with initial metal ion concentrations of 25, 50, 75, 100, and 150 mg L^{-1}. An example of the data collected is shown in Table 2 for the Zn(II)PMA sorption of Pb(II) ions.

The experimental data was plotted and fit to the models to determine the model parameters and the correlation factor. The plots of the experimental values for the Zn(II)PMA sorption and the line fits and correlation for both Langmuir and Freundlich models are shown in Figure 13. The correlation factors for the Langmuir ($R^2 = 0.9828$) and Freundlich ($R^2 = 0.9934$) models indicate some heterogeneous, Freundlich-type adsorption. Fitting the experimental data from the Cu(II)PMA adsorption of Pb(II) results in correlation factors for Langmuir ($R^2 = 0.978$) and Freundlich ($R^2 = 0.808$) which indicate a homogeneous, Langmuir-type adsorption. The Fe(II)PMA adsorption of Cr(VI) fits the Freundlich model ($R^2 = 0.9562$) much better than the Langmuir model ($R^2 = 0.5773$), indicating heterogeneous adsorption. However, the Fe(III)PMA adsorption of Cr(VI) correlates to both Freundlich ($R^2 = 0.9602$) and Langmuir ($R^2 = 0.9668$) models, indicating some heterogeneous binding.

Table IV Experimental data from the Zn(II)PMA adsorption

Pb(II) initial concentration $(C_o\,/\,mg\,L^{-1})$	Pb(II) equilibrium concentration $(C_e\,/\,mg\,L^{-1})$
25	1.5
50	8.69
75	16.94
100	32.4
150	65

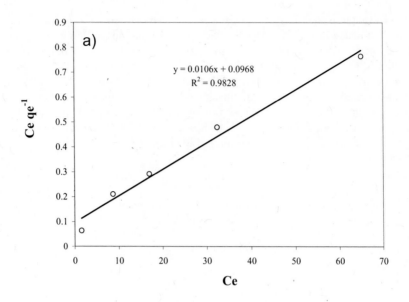

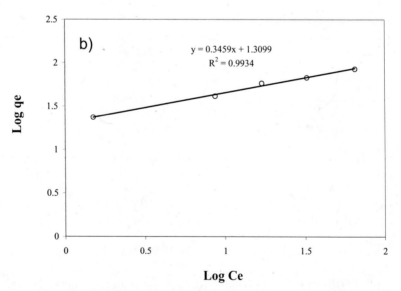

Figure 13. Linearized isotherm of Zn(II)PMA after the contact with a aqueous solution of Pb a) Langmuir model and b) Freundlich model. (Reproduced from reference 1. Copyright 2007 American Chemical Society)

SEM and EDX Elemental Analysis of the Polymer Surface after Sorption

The elemental composition of the surfaces of the polymers were analyzed after the sorption experiments with EDX analysis in the SEM. The elemental composition was compared to that of the original polymer, as shown in Figure 14 for Zn(II)PMA, to confirm surface adsorption of the Pb(II) or Cr(VI) metal ions.

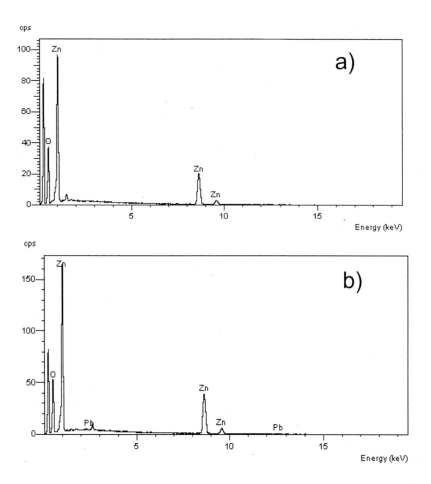

Figure 14. EDS spectra of Zn(II)PMA before (a) and after (b) contact with the aqueous lead solution. (Reproduced from reference 1. Copyright 2007 American Chemical Society)

X-Ray Photoelectron Spectroscopy (XPS)

XPS was done on the Zn(II)PMA after sorption. XPS is used to identify the interaction of a metal ion with surface chemical groups on an adsorbent. Interactions between a metal ion and an atom on the surface of the adsorbent changes the distribution of the electrons around the corresponding atoms - electron-donating ligands can lower the binding energy (BE) of the core electrons, while electron-withdrawing ligands can increase it. The XPS spectra of the Zn(II)PMA after the contact with lead ions in aqueous solution is shown in Figure 15. The spectrum confirms that carbon, oxygen, zinc and lead atoms are on the surface of the polymer. A BE value of 530 eV is usually attributed to the oxygen in the C=O and OH groups. These groups may be involved in the adsorption of lead ions. The BEs around 138 eV may be assigned to the bond of Pb-O. The results suggest that the carboxyl and hydroxyl groups are involved in binding the lead ions.

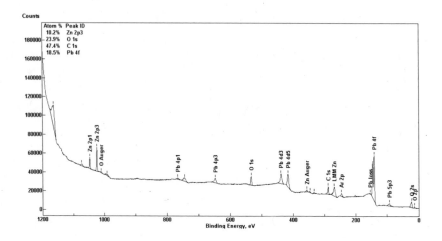

Figure 15. XPS spectra of the Zn(II)PMA after the adsorption of lead ions. (Reproduced from reference 1. Copyright 2007 American Chemical Society)

Lead sorption mechanism

In order to explain the lead removal from aqueous solution it should be mentioned the relevance of carboxyl group for metal ion binding. In previous metal ion sorption studies, the blocking carboxyl groups of algal species by esterification decreased the binding capacity of Cu and Al(30). This decrease was correlated to the degree of esterification. Other studies indicate that the blocking with propylene oxide of weakly acidic groups (pK near 5) in *S. fluitans*, which are likely to be carboxyl groups, was 90% effective and resulted in 80% reduction of the metal ion binding (31). This demonstrated the responsibility of weakly acidic groups for most of the metal ion binding.

Since no metal from the polymer (zinc, copper, or iron) was detected in the treated aqueous solutions, a possible ion exchange mechanism is not considered in these sorption processes. Since ion exchange is not a mechanism in these studies, it is reasonable to believe that the metal ions (lead or chromium) are binding on free sorbent sites.

Comparison with other sorbents

An accurate direct comparison of Zn(II)PMA, Cu(II)PMA, Fe(II)PMA, and Fe(III)PMA with other sorbent materials is not feasible owing to different applied experimental conditions. However, for the sake of qualitative comparison, reported values for other lead and hexavalent chromium sorbents are listed in Tables V and VI.

Table V Comparison of Pb(II) sorption capacity of different materials.

Adsorbent	Reported sorption capacities (mg g^{-1})	Reference
Amine-PAN fibers	76.12	(33)
Treated Opuntia	51.82	(22)
Zn(II)PMA	**20.11**	**(1)**
Rice hulls	11.40	(18)
Undyed sawdust	7.3	(11)
Cu(II)PMA	**6.22**	**(2)**
Bentonite	6	(8)
Wollastonite	0.217	(9)

Table VI Comparison of Cr(VI) sorption capacity of different materials.

Adsorbent	Reported sorption capacities (mg g^{-1})	Reference
Fe(II)PMA	**22.26**	**(3)**
Polyacrylamide-sawdust	12.4	(24)
Pyrolytic Ashes	6.45	(19)
Opuntia	6.22	(15)
Granular activated carbon	5.09	(6)
Fe(III)PMA	**3.87**	**(3)**
Typha Latifolia	3.69	(19)
Activated bagasse carbon	0.19	(32)

Conclusions

In the lead sorption studies, Zn(II)PMA exhibited heterogeneous Freundlich behavior adsorbing 94% of the lead in a 25 ppm solution and 67% in a 150 ppm solution with a capacity of 20.11 mg per gram of sorbent, while Cu(II)PMA matched better with the homogeneous Langmuir model adsorbing 56% from a

75 ppm solution and 48% from a 150 ppm solution with a capacity of 6.22 mg per gram of sorbent. The Zn(II)PMA has a higher capacity than many adsorbents used in wastewater treatment. In the Cr(VI) oxyanion studies, Fe(II)PMA exhibited a heterogeneous Freundlich behavior adsorbing 70% of the chromate ions in a 75 ppm solution with a capacity of 22.26 mg per gram of sorbent, while Fe(III)PMA behaved in a homogeneous Langmuir mechanism adsorbing only 30% of a 75 ppm solution with a capacity of 3.52 mg per gram. The Fe(II)PMA has a high capacity for chromate ions and shows great promise as an effective sorbent.

Acknowledgements

The authors wish to acknowledge the support given by CONACYT and The Universidad Autonoma del Estado de Mexico, especially the Facultad de Química (Project 2228/2006).

References

1. Ureña-Nuñez, F.; Barrera-Díaz, C.; Bilyeu, B. Gamma Radiation-Polymerized Zn(II) Methacrylate as a Sorbent for Removal of Pb(II) Ions from Wastewater. *Ind. Eng. Chem. Res.* **2007**, *46*, 3382.
2. Barrera-Díaz, C.; Palomar-Pardavé, M.; Romero-Romo, M.; Ureña-Nuñez, F. Lead Removal from Wastewater Using Cu(II) Polymethacrylate Formed by Gamma Radiation. *J. Polym. Res.* **2005**, *12*, 421.
3. Ureña-Nuñez, F.; Díaz-Jiménez, P.; Barrera-Díaz, C.; Romero-Romo, M.; Palomar-Pardavé, M.; Gamma radiation-induced polymerization of Fe(II) and Fe(III) methacrylates for Cr(VI) renoval from wastewater. *Rad. Phys. Chem.* **2003**, *68*, 819.
4. Barrera-Díaz, C.; Palomar-Pardavé, M.; Romero-Romo, M.; Martínez, S. Chemical and electrochemical considerations on the removal process of hexavalent chromium from aqueous media. *J. Appl. Electrochem.* **2003**, *33*, 61.
5. Faust, S. D.; Aly, O. M. *Adsorption Process for Water Treatment*; Butterworth: Stoneham, MA, 1987.
6. Ramos, R. L.; Martínez, A. J.; Coronado, R.M.G. Adsorption of chromium (VI) from aqueous solutions on activated carbon. *Water Sci. Technol.* **1994**, *30*, 187.
7. Dimitrova, S. V. Use of granular slag columns for lead removal. *Water Res.* **2002**, *36*, 4001.
8. Khan, S. A.; Riaz-ur-Rehman, A.; Khan, M. A. Adsorption of chromium (III), chromium (VI) and silver (I) on bentonite. *Waste Manage.* **1995**, *15*, 271.
9. Yadava, K. P.; Tyagi, B. S.; Singh, V. N. Effect of temperature on the removal of lead(II) by adsorption on China clay and wollastonite. *J. Chem. Tech, Biotech.* **1991**, *51*, 47.

10. Stana-Kleinschek, K.; Strand, S.; Ribitsch, V. Surface Characterization and Adsorption Abilities of Cellulose Fibers. *Polym. Eng. Sci.* **1999**, *39*, 1412.

11. Shukla, S. R.; Sakhardande, V. D. Column studies on metal ion removal by dyed cellulosic materials. *J. Appl. Polymer Sci.* **1992**, *44*, 903.

12. Boduu, V. M.; Abburi, K.; Talbott, J. L.; Smith, E. D. Removal of hexavalent chromium from wastewater using new composite chitosan biosorbent, *Environ. Sci. Tech.* **2003**, *37*, 4449.

13. Volesky, B. *Sorption and Biosorption*; BV-Sorbex: St. Lambert, Quebec, 1987.

14. Barrera-Díaz, C.; Roa-Morales, G.; Bilyeu, B. Identification and quantification techniques in the removal of heavy metals from aqueous solutions using low cost sorbents. *Applications of Analytical Chemistry in Environmental Research*; Palomar, M., Ed.; Research Signpost; India, 2005; p145.

15. Barrera, H.; Ureña-Nuñez, F.; Bilyeu, B. ; Barrera-Díaz, C. Removal of chromium and toxic ions present in mine drainage by Ectodermis of Opuntia, *J. Haz. Mat.* **2006**, *B136*, 846.

16. Brown, P. A.; Gill, S. A.; Allen, S. J. Metal removal from wastewater using peat. *Water Res.* **2000**, *34*, 3907.

17. Kratochvil, D.; Pimentel, P.; Volesky, B. Removal of trivalent and hexavalent chromium by seaweed biosorbent. *Environ. Sci. Technol.* **1998**, *32*, 2693.

18. Roy, D.; Greenlaw, P. N.; Shane, B. S. Adsorption of heavy metals by green algae and ground rice hulls. *J. Environ. Sci. Health.* **1993**, *28*, 37.

19. Barrera-Díaz C., Colín-Cruz A., Ureña-Nuñez F., Romero-Romo M, Palomar-Pardavé M. Cr(VI) removal from wastewater using low cost sorbent materials: roots of typha latifolia and ashes. *Environ Technol*, **2004**, 25, 907-917.

20. Nasernejad, B.; Zadeh, T. E.; Bonakdar, P. B., Bygi, M. E.; Zamani, A. Comparison for biosorption modeling by heavy metals (Cr(III), Cu(II), Zn(II)) adsorption from wastewater by carrot residues. *Process Biochem.*, **2005**, *40*, 1319.

21. Kiefer, R.; Höll, W. H. Sorption of heavy metals onto selective ion-exchange resins with aminophosphonate functional groups. *Ind. Eng. Chem. Res.* **2001**, *40*, 4570.

22. Bernal-Martínez, L. A.; Hernández-Lopez, S.; Barrera-Díaz, C.; Ureña-Nuñez, F.; Bilyeu, B. Pb(II) Sorption under Batch and Continuous Mode Using Natural, Pretreated, and Amino-Modified Ectodermis of Opuntia. *Ind. Eng. Chem. Res.* **2008**, *47*, 1026.

23. Linares-Hernández, I.; Barrera-Díaz, C.; Roa-Morales, G.; Bilyeu, B.; Ureña-Nuñez, F. A combined electrocoagulation-sorption process applied to mixed industrial wastewater. *J. Haz. Mat.* **2007**, *144*, 240.

24. Raji, C.; Anirudhan, T. S. Batch Cr(VI) removal by polyacrylamide-grafted sawdust: kinetics and thermodynamics. *Water Res.*, **1998**, *32*, 3772.

25. Unnithan, M. R.; Vinod, V. P.; Anirudhan T. S. (2004) Synthesis, characterization and application as a chromium (VI) adsorbent of amide-modified polyacrylamide-grafted coconut coir pith. *Ind. Eng. Chem. Res.* **2004**, *43*, 2247.

94

26. Deng, S.; Bai, R.; Chen, J. P. Aminated polyacrylonitrile fibers for lead and copper removal. *Langmuir.* **2003**, *19*, 5058.
27. Takács, E.; Wojnárovits, L.; Borsa, J., Papp, J.; Hargittai, P.; Korecz, L. Modification of cotton-cellulose by preirradiation grafting *Nucl. Instr. Meth. In Phys. Res. B*, **2005**, *236*, 259.
28. Galván-Sánchez, A. F.; Ureña-Nuñez, F.; Flores-Llamas, R.; López-Castañares, R. Determination of the crystalinity index of iron polymethacrylate. *J. Appl. Polym. Sci.* **1999**, *74*, 995.
29. American Public Health Association/American Water Works Association/Water Environment Federation Standard Methods for the Examination of Water and Wastewater, 20th Ed., Washington DC, USA 1998.
30. Schiewer, S.; Volesky, B. Modeling of the proton-metal ion exchange in biosorption. *Environ. Sci. Technol.* **1995**, *29*, 3049.
31. Fourest, E.; Volesky, B. Contribution of sulfonate groups and alginate to heavy metal biosorption by the dry biomass of Sargassum fluitans. *Environ. Sci. Technol.* **1996**, *30*, 277.
32. Chand, S.; Agarwal, V. K.; Pavakumar, C. Removal of hexavalent chromium from wastewater by adsorption. *India J. Envir. Health* **1994**, *36*, 151.
33. Shubo, D.; Renbi, B.; Paul, C. J. Aminated Polyacrylonitrile Fibers for Lead and Copper Removal. *Langmuir* **2003**, *19*, 5058.

Chapter 6

Copper Ion Exchange Studies of Local Zeolitic Tuffs

Özge Can[1,2], Devrim Balköse[1], Semra Ülkü[1]

[1]Department of Chemical Engineering, Izmir Institute of Technology, Urla, Izmir, Turkey
[2]Current Address: Department of Chemical and Biomedical Engineering, Cleveland State University, Cleveland, Ohio 44115

Copper(II) was selected as a model ion and Cu^{2+} exchange batch and column studies were made using natural zeolite samples having different clinoptilolite contents. 8.33 and 10.00 mg/g copper ion exchange capacities were determined for zeolites with 64% (CP2) and 80% (CP1) clinoptilolite contents, respectively. It has quantitatively been shown that Cu^{+2} ions were exchanged with Na^+, K^+, Ca^{2+} and Mg^{2+} ions. Flow rate had the most important effect on breakthrough point rather than column height and inlet solution concentration for column studies.

Industrial waste waters frequently contain high levels of heavy metals and efficient treatment is needed as it is crucial to avoid water pollution due to increasing social and economic importance of environmental conservation. Heavy metal pollution often results from industrial use of organic compounds containing metal additives in the petroleum and organic chemical industries. Clinoptilolite is probably the most abundant zeolite in nature. Characteristic clinoptilolite rocks consist of 60-90% clinoptilolite and other constituents are often being mainly feldspars, clays, glass, and quartz. Ion exchange is considered to be one of the most cost effective methods if low cost ion exchangers such as natural zeolites are used. There are many studies related to equilibrium and kinetics of cation exchange on natural clinoptilolite *(1-3)*. Removal of Ni(II) ions *(4,5)*, Cu(II) ions *(6-8)*, Cd(II), Ni(II) and Cu(II) *(9)*, Ni(II), Cu(II), Pb(II) and Cd(II) *(10)* by clinoptilolite minerals from different origins has been investigated by previous researchers. Column dynamics of heavy metal removal by clinoptilolite has been investigated previously by other workers *(6, 11, 12)*.

Since ion exchange capacities depend on the clinoptilolite content of the zeolitic tuffs, quantitative purity determination is important. Quantitative determination of phase abundances using X-ray diffraction (XRD) is a widely used technique *(3-16)*. Among the various methods, Reference Intensity Ratio (RIR) method is one of the most common because it can provide reliable results for all sample types. RIR method requires the measurement of intensities of specific reflections of each phase in standards and samples. RIR is defined as the intensity of the peak of interest for a given phase divided by the intensity of a peak from a standard in a 50:50 mixture *(17)*.

In this study, investigation of copper exchange performance of zeolitic tuffs from Gördes/Manisa, Turkey was aimed to be studied taking their zeolitic content into consideration. To achieve this goal, zeolitic content of the tuffs was determined by XRD.

Experimental

In this study, clinoptilolite rich natural zeolite samples called as CP1 and CP2 taken from different deposites of Gördes/Manisa region were used. The zeolitic content of CP1 was determined in the study of Top and Ülkü *(18)* as 80% and purity of CP2 was determined in this study. Since the zeolite mineral in these samples had been identified as clinoptilolite in previous studies using chemical, thermal gravimetric analysis, differential scanning calorimetry, and Fourier transform infrared spectroscopy *(19, 20)*, only the purity was determined in the present study.

Sample preparation for X-ray diffraction analysis

Natural zeolite samples were taken representatively from Gördes/Manisa region, Enli Mining, corp. Approximately 500 kg samples were collected and then simultaneously divided into four groups until the desired amount of samples were obtained. These samples were crushed into 1-2 cm range of particle size and then into < 2 mm size range with a jaw crusher (Fritsch, Pulverisette 1). Particles smaller than 2 mm size were then wet sieved into five groups such as; > 1.7 mm, 425 μm – 1.7 mm, 106 – 425 μm, 25 – 106 μm and finally smaller than 25 μm. The reason for applying wet sieving was to prevent the smaller particles to stick on the larger ones and hence change the ion exchange capacity of these particles. After the wet sieving process all zeolite fractions were dried at 105 °C overnight. In further characterization studies all fractions were ground under 25 μm (Fritsch, Pulverisette 9) to eliminate the effect of particle size in x-ray diffraction analysis.

Optimum grinding time

Less than 25 µm particles are required for a well resolved X-ray diffraction diagram. In order to determine the optimum grinding time to reduce the particles under 10 µm, the same amount of samples (30 g) were ground for 1, 2 and 4 minutes and then the particle size distribution of ground products was determined by using a sedigraph (Micromeritics, Sedigraph 5100). For this purpose 1 gram of sample obtained with different grinding periods was dispersed in 50 cm^3 of 50 w% sodium meta phosphate solutions by stirring in an ultrasonic bath (Ultrasonic LC30). These even dispersions were used in particle size distribution determination.

Quantitative analysis of clinoptilolite

Standard materials used for quantitative analyses were obtained from Mineral Research, Clarkson, New York with +90 % pure, clinoptilolite, IDA, (# 27031, Castle Creek, Idaho) and CAL, (# 27023, Hector, California). RIR standard mixtures were prepared by mixing 1.0 µm α-alumina powder (corundum) to each mineral standard in a 50:50 ratio by weight. Mixtures of the zeolitic tuff sample and corundum were also prepared in the same manner. XRD data were collected on X-ray diffractometer (Philips X'pert Pro) employing the Cu Kα radiation of power settings of 30 kV and 30 mA. Data were collected using a step size of 0.02° 2θ and a count time of 2 s/step. Several discrete 2θ ranges between 5 and 40° 2θ were used to measure only the reflections of interest. In order to determine the RIR values for clinoptilolite, six replicate XRD scans were conducted on each RIR standard. Before each replicate analysis, the standard was removed from the sample mount, remixed with the standard remaining in the sample bottle, and the sample was remounted. Two RIR standards were prepared with six replicate runs conducted on each as outlined above. The reference intensity ratio is defined as the intensity of the peak of interest for a given phase divided by the intensity of a peak from a standard in a 50:50 mixture *(17)*.

Samples for Ion Exchange Studies

Clinoptilolite samples having 0.85 -2.0 mm particle size range were used in the experiments in order to prevent small particles from being carried over within the liquid phase, to minimize pressure drop, to prevent axial dispersion and to provide sufficient mass transfer area. These samples were obtained by dry sieving of crushed samples having < 2 mm particle size that were crushed using a jaw crusher (Fritsch, Pulverisette 1). The fractionated clinoptilolite samples were dried in the oven at 110 °C and kept in desiccator until needed for ion exchange experiments without any further pretreatment.

Batch Kinetic and Equilibrium Studies

Equilibrium behavior of the system and the effect of initial concentration on percent removal were investigated for both CP1 and CP2. 100 ml of copper nitrate solutions with different concentrations were prepared. Then 1 g of clinoptilolite sample was placed in volumetric flasks and shaken at 130 rpm, 29 °C in water bath. For five days long, samples were taken at specific time intervals until no further metal uptake from zeolite samples was observed. Finally, samples were analyzed using Inductively Coupled Plasma- Atomic Emission Spectrometry (Axial ICP-OES 96, Varian Liberty Series). 1 % (w/v) of nitric acid added to the solutions to prevent precipitation of metal hydroxides and also to eliminate the matrix effect during ICP analyses. The pH of the solutions during ion exchange was measured using Methrohm 744 pH meter.

Packed Column Studies

In fixed bed studies, ion exchange experiments were carried out at 29 ± 1 °C in a pyrex ion exchange column of 30 cm height and 1.5 cm inner diameter. The feed was introduced to the system using a diaphragm pump (Cole- Parmer) in up flow mode. In these experiments, samples were collected at specific time intervals by using an automatic fraction collector (Atto Biocollector, AC-5750). Effluent sample concentrations were determined using ICP-OES. The experiment numbers and the operating conditions of columns experiments are given in Table I. Effects of feed concentration (100, 159, 200 mgdm^{-3}), packing height (12.5, 18.8 and 25.0 cm) and flow rate (1.87, 2.86, 3.6, 4.8 cm^{3}min^{-1}) on breakthrough curves were investigated. The breakthrough time was taken as the time when the outlet solution concentration reached to 5 mgdm^{-3}.

Table I. Operating Conditions of Column Experiments

Experiment No	Zeolite type	Feed concentration $mg\,dm^{-3}$	Flow rate cm^3min^{-1}	Bed height cm	Amount of zeolite g
1	CP1	100	4.75	18.8	19.95
2	CP1	159	4.75	18.8	19.95
3	CP1	200	4.75	18.8	19.95
4	CP1	200	1.87	25.0	26.75
5	CP1	200	3.60	25.0	26.75
6	CP1	200	4.75	25.0	26.75
8	CP1	200	4.75	12.5	14.06
9	CP1	200	4.75	18.8	19.95
10	CP1	200	4.75	25.0	26.75
11	CP2	200	1.87	25.0	26.75

Dried clinoptilolite samples before and after ion exchange were analyzed by EDX using SEM, Philips XL 30S FEG. During EDX analyses data taken from 10 different parts of the samples were evaluated to determine the standard deviations. Representative zeolite particles were cut into two pieces and the fracture surfaces were analyzed at points close to their external surface and at their center for Cu, K, Na, Ca, and Mg elements to study the exchange of ions.

Results and Discussion

Representative clinoptilolite crystals are clearly seen for CP1 and CP2 zeolites in SEM micrographs taken at high magnification in Figure 1. The average size of the crystals is ranging between 5 to 10 μm.

(a)

(b)

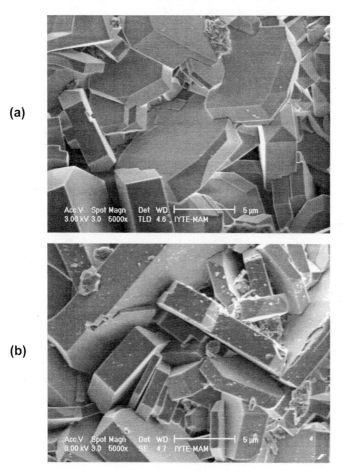

Figure 1. SEM micrographs of (a) CP1 zeolite crystals (b) CP2 zeolite crystals.

Determination of the optimum grinding time

Size distribution of the samples ground for different periods are seen in Figure 2. It is seen that 80% of the particles are under 8, 12, and 17 µm for 4, 2 and 1 minute grinding periods, respectively. Particles obtained for 2 minutes grinding time were chosen for quantitative analysis, since most of the particles were less than 10 µm which is necessary for well defined diffraction peaks.

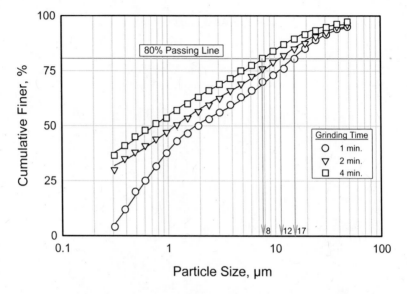

Figure 2. Particle size distribution of zeolite particles ground for different periods.

Purity determination

Generally, it is more desirable to use the sum of reflections from a localized region in the diffraction pattern than to use individual reflections from a phase. Thus, for clinoptilolite, two separate intensity regions were chosen: the 020 reflection at 9.8° 2θ and the sum of the intensity in the range 22.1-23.0° 2θ (Figure 3).

It has been shown that precise RIR values for a reflection or sum peak can be obtained by plotting the RIR for a given peak versus the ratio of the intensities measured for this peak to a second peak or reflection *(17)*. This technique minimizes the effects of both preferred orientation and chemical variability by producing the curve of RIR values. Equation 1 gives the relation between the RIR values for the 020 clinoptilolite reflection (I_{020}) and the ratio of the intensity of 020 reflection and the (22.1-23 °2θ) sum peak for the clinoptilolite standards ($\Sigma I_{22.1-23}$):

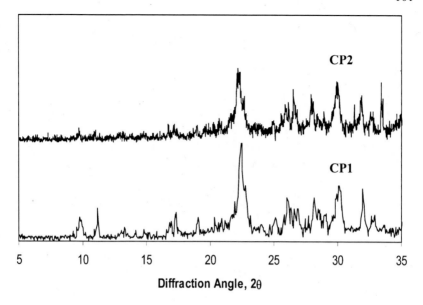

Figure 3. XRD spectra of the CP1 and CP2 zeolites.

$$RIR_{ur} = 0.54\left(\frac{I_{020}}{\Sigma I_{22.1-23^o}}\right) + 0.22 \tag{1}$$

It is important to note that RIR values are dependent on the instrument geometry and physical dimensions of the sample holder. Therefore, RIR values should be measured for each instrument/sample configuration *(17)*.

The weight fraction of the clinoptilolite phase in the unkown sample (X_u) is given by:

$$X_u = \frac{I_{u(020)}}{I_{r(020)}}\frac{X_r}{RIR_{ur}} \tag{2}$$

where X_r is the weight fraction of clinoptilolite in the reference.

The clinoptilolite contents of CP1 and CP2 zeolites were calculated from equation 2 as 80 and 64 ± 5 %, respectively.

Ion Exchange

Equilibrium behavior of the systems was investigated for both CP1 and CP2. Although the amount of copper removed from the solution increased, the percent removal of the copper ions decreased for both zeolites by increasing the initial copper ion concentration. It is clearly seen in Figure 4 that, the percent removal of Cu^{2+} ions from the solution was greater for CP1 zeolite for all initial ion concentrations. The higher the Cu^{+2} concentration in solution the lesser is the removal efficiency for both zeolites. At a copper concentration of 50 mgdm^{-3}, removal efficiencies of CP1 and CP2 zeolites were achieved as 88 and 81 %, respectively. At higher copper concentration of 200 mg/dm^3, the removal efficiency decreased to a value of 50 and 42 % for CP1 and CP2 zeolites, respectively. Similar observations were found by Türkmen *(21)*. For 87 μm sized particles 30 and 17% removal were observed for initial Cu^{+2} concentrations of 29 and 346 mg/dm^3, respectively *(21)*.

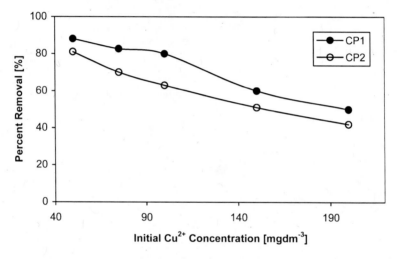

Figure 4. Percent removal with respect to initial Cu^{2+} concentration, at 30°C for 1 g zeolite per 100 cm^3 solution.

Since ion exchange is a stoichiometric reaction, total equivalents of exchanging cations between the solution and the ion exchanger should be equal to each other. Therefore, a mole balance was constructed between the major exchangeable cations initially present in the zeolite structure and the copper ions in the solution. In Table II, the equilibrium ion concentrations for Cu, Na, K, Ca and Mg elements in solutions are reported for CP1 and CP2 zeolites. In Table III the amount of Cu^{2+} ions transferred to the solid phase was determined by the difference in the change in milliequivalents of Cu^{2+} ions and by summing up the changes in milliequivalents of Na^+, K^+, Mg^{2+} and Ca^{2+} ions. The percent differences of the exchanged ions, calculated by the two methods, were generally within the tolerable limits for both zeolites.

Ion exchange isotherms of copper were constructed for CP1 and CP2 zeolites with corresponding Langmuir model in terms of species concentration in the solid phase as a function of its value in solution (Figure 5). Langmuir model described the equilibrium behavior of both systems quite well. It can be seen from Figure 5 that natural zeolite containing 80% clinoptilolite, CP1, gave rise to a better ion exchange capacity for copper than CP2 zeolite containing 64% clinoptilolite. Ion exchange capacities were determined as 10.00 mg (0.32 meq) Cu^{2+} / g CP1 zeolite and 8.33 mg (0.26 meq) Cu^{2+} / g CP2 zeolite. The Langmuir coefficients were found as 0.12 and 0.05 for CP1 and CP2, respectively.

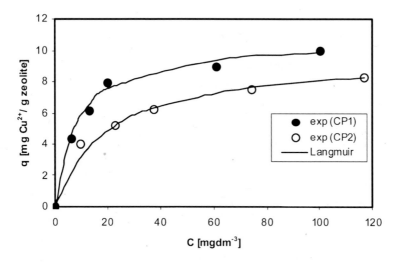

Figure 5. Equilibrium uptake isotherms of copper with CP1 and CP2 zeolite samples and corresponding Langmuir model.

Table II. Ion concentrations in mg dm^{-3} at equilibrium for different initial Cu^{2+} concentrations in aqueous phase for CP1 and CP2 zeolites

$C_o (Cu^{2+})$ [$mgdm^{-3}$]	Equilibrium concentrations for CP1 zeolite					Equilibrium concentrations for CP2 zeolite				
Cu^{+2}	Cu^{2+}	Na^+	K^+	Ca^{2+}	Mg^{2+}	Cu^{+2}	Na^+	K^+	Ca^{2+}	Mg^{2+}
50	6	13	12	8	2	10	8	9	6	8
75	14	15	11	11	2	26	8	13	10	13
100	21	22	11	12	2	33	8	10	9	11
150	61	20	16	27	3	75	8	12	13	15
200	101	29	22	23	-	116	9	15	16	18

C_o: Inlet ion concentration in solution

Table III. Equilibrium uptake data for copper exchange of clinoptilolite samples for Cu^{2+} ions

C_o (Cu^{2+}) [mgdm^{-3}]	CP1			CP2		
	q_{eq}^a [meq/g]	q_{eq}^b [meq/g]	% difference	q_{eq}^a [meq/g]	q_{eq}^b [meq/g]	% difference
50	0.140	0.146	4.07	0.127	0.147	-15.75
75	0.193	0.179	7.21	0.165	0.166	-0.61
100	0.249	0.238	4.73	0.197	0.189	4.06
150	0.289	0.293	1.42	0.238	0.261	-9.66
200	0.319	0.303	5.26	0.262	0.312	-19.08

a: from Cu^{2+} measurements directly

b: from K$^+$+Na$^+$+Ca^{2+}+Mg^{2+} measurements

The ion exchange capacities for Cu^{2+} determined in this work fall within the corresponding range reported in the literature (22). On the contrary, in the studies performed by Cincotti et al. (23) and Türkmen (21) the ion exchange capacities determined by using natural clinoptilolite were much lower than the values indicated in this study. On the other hand, in the studies performed by Nikashina et al. (24), Guangsheng et al. (25), Langella et al. (26), ion exchange capacities of sodium form of clinoptilolite were higher. The reason for obtaining such diverse results is using clinoptilolites having different purity levels, compositions, different forms and particle sizes and most importantly, they are from different origins.

Kinetics of Ion Exchange

There were simultaneous change in solution pH and Cu(II) ion concentration in solution in contact with the zeolitic tuff as seen in Figure 6 and 7. In Figure 6, Cu(II) exchange was nearly completed in the first 100 minute period and fraction exchanged versus squareroot of time showed linear behavior indicating diffusion controlled mass transfer. Using Equation 3 and average of the slopes of the initial linear region of the curves in Figure 6, the apparent diffusivity, D, of Cu(II) ions was found to be 3.1×10^{-10} m^2/s for the average particle size (r) of 1.4 mm.

$$M_t/M_\infty = 6/r(Dt/\pi)^{1/2} \tag{3}$$

Where M_t and M_∞ are the amount exchanged at time t and at equilibrium, respectively.

While kinetics of ion exchange fitted to pseudo second order relation with respect to difference in equilibrium and instantaneous metal ion concentration in solid phase by previous workers (9), solid diffusion model fitted better to the experimental data of the present study.

As seen in Figure 7, pH of the copper ion solutions in contact with zeolitic tuff particles increased with respect to time. pH changed from 4.6 to 5.8 for 200 mgdm^{-3} Cu^{2+} and from 4.2 to 5.4 for 50 mgdm^{-3} in 2800 minutes time period. This corresponds to nearly 0.06 moldm^{-3} decrease in hydrogen ion concentration. The basic groups in zeolitic tuff neutralized the protons in solution. On the other hand, when there was KCl dissolved in solution, no change in solution pH was observed by Doula et al. *(8)*.

pH change was slower than Cu^{2+} concentration change as seen in Figure 6 and 7 indicating that these changes occur by different mechanisms.

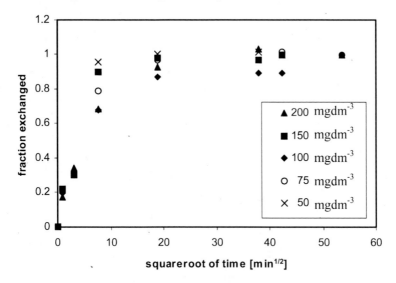

Figure 6. Fractional uptake of Cu^{2+} for different initial concentrations for CP1 zeolite with respect to squareroot of time.

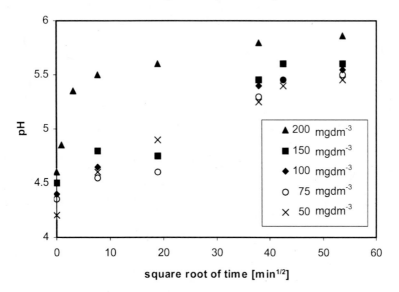

Figure 7. Change of pH with respect to squareroot of time for CP1 zeolite.

Distribution of ions in zeolitic tuff particles

Representative CP1 zeolite samples were cut into two parts to see how the copper ions moved towards the inner volume during ion exchange. EDX concentrations of only exchangeable cations namely, Na^+, K^+, Mg^{2+}, Ca^{2+} and Cu^{2+} before and after ion exchange are represented by a bar graph in Figure 8. Each bar represents the average results of EDX analyses of 10 points taken near the external surface or center of the particle before and after ion exchange. It is seen that after copper exchange process, weight percentages of Na^+, K^+, Mg^{2+} and Ca^{2+} ions were decreased, while weight percent of Cu^{2+} ions was increased. While Na^+, K^+, Mg^{2+}, Ca^{2+} concentrations were higher at the center of the particles, Cu^{2+} concentration was higher near the surface for ion exchanged particle indicating slow diffusion of ions in the solid phase. Thus, even if the solution concentration did not change at the pseudo equilibrium state in batch and column studies, equilibrium has not been reached with respect to solid phase.

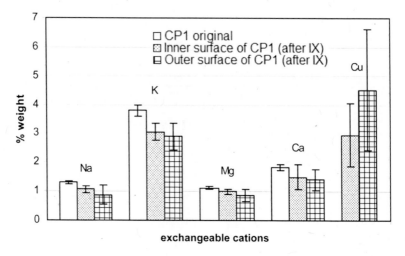

Figure 8. Composition of CP1 zeolite, before and after ion exchange process for 200 mgdm^{-3} feed concentration, 1.87 cm^{3}min^{-1} flow rate and 25 cm packing height.

Column Dynamics

Breakthrough curves were constructed and ion exchange capacities were calculated for each run. C represents the time-dependent outlet metal concentration of the bed. Ion exchange capacities of the zeolites were compared for the packing height, flow rate and initial copper concentration of 25 cm, 1.87 cm^3/min and 200 mgdm^{-3}, respectively (Figure 9). It was determined that the breakthrough time for both zeolites was 380 min. The system has reached equilibrium at around 2500 and 4000 min for CP2 and CP1. Although the breakthrough curves were generally close to each other, CP1 showed higher copper exchange level within the time interval of 1000-4000 min.

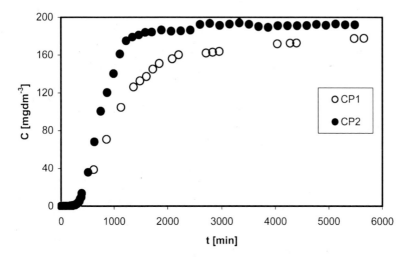

Figure 9. Comparison of the breakthrough curves for Cu^{2+} exchange on CP1 and CP2 zeolites with respect to relative concentration and time at 30°C, 1.87 cm^3min^{-1}, 12.5cm packing height and feed concentration of 200 mgdm $^{-3}$.

Table IV. Inlet and outlet pH values and breakthrough times for different experimental runs

Experiment No	Initial pH	Final pH	Breakthrough time [minutes]
1	4.4	5.8	150
2	4.5	5.9	120
3	4.6	5.9	30
4	4.5	5.9	360
5	4.8	5.9	186
6	4.7	5.9	40
7	4.8	5.3	55
8	4.6	5.9	30
9	4.7	5.9	40
10	4.5	5.6	360

When the batch studies were compared with the column studies, in batch experiments, copper removed from the solution was determined as 10 mg and 8.33 mg for 1 g of CP1 and CP2 zeolites, respectively. In column studies, at the same initial Cu^{2+} concentration of 200 mgdm^{-3}, copper exchange capacities were calculated from breakthrough curves as 9.46 mg Cu^{2+}/g CP1 and 8.04 mg Cu^{2+}/g CP2 indicating that the column experiments yield ion exchange capacities close to the equilibrium values which are generally more reliable.

The effects of flow rate, inlet concentration and packing height on breakthrough curves are shown in Figures 10, 11 and 12, respectively. In the interpretation of the breakthrough curves, the breakthrough point was set at 5 mg dm^{-3} outlet concentration. The breakthrough times reported in Table IV for different experimental runs showed that generally, the lower the inlet concentration, flow rate and the higher the packing height, the longer the breakthrough time. Flow rate, rather than the packing height and inlet concentration had the most important effect in breakthrough time. In a previous study (6), flow rate similarly affected the breakthrough time.

The pH values at the exit flow were higher then the inlet at the end period of the column dynamics experiments as seen in Table IV, indicating the neutralizing action of the clinoptilolite used in the study. On the contrary, with a different zeolitic tuff in a longer column, the pH was lowered through the column (12). This was explained by two opposing effects: hydrolysis of metal ions and neutralization of protons by zeolite (12).

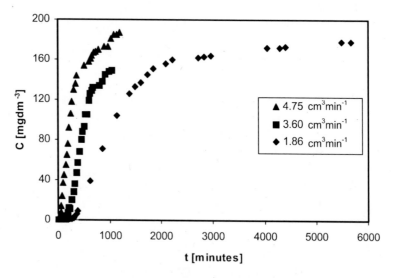

Figure 10. Effect of flow rate on breakthrough curves for CP1 zeolite (packing height 25 cm and feed concentration 200 mgdm^{-3}).

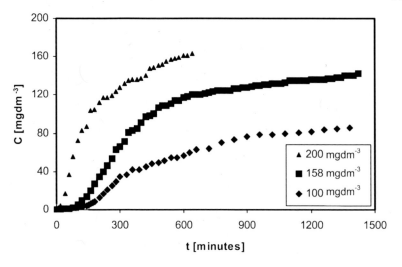

Figure 11. Effect of feed concentration in mg/dm³ on breakthrough curves for CP1 zeolite (flow rate 4.75 cm³min⁻¹ and packing height 18.75 cm).

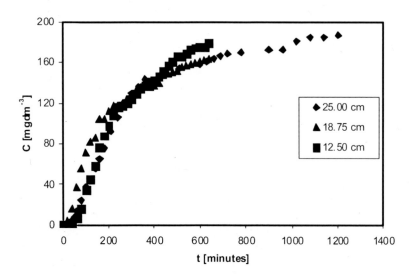

Figure 12. Effect of packing height on breakthrough curves for CP1 Zeolite (flow rate 4.75 cm³/min and feed concentration 200 mgdm⁻³).

110

Summary

The percent removal of copper ions from aqueous solutions was greater for the zeolitic tuff having higher clinoptilolite content. Copper ion exchange capacities were determined as 10.00 mg / g and 8.33 mg / g for CP1 and CP2 zeolites having 80 and 64 % clinoptilolite, respectively. Langmuir model described the equilibrium behavior of the systems quite well. It has quantitatively been shown that Cu^{2+} ions were exchanged with Na^+, K^+, Ca^{2+} and Mg^{2+} ions. Higher Cu^{2+} concentrations at the external surface than the center of the copper ion exchanged particles indicated slow diffusion of the ions in solid phase. Flow rate was the most important variable effecting breakthrough times from the column. The equilibrium, kinetic and column dynamic data obtained in this study could be used to design local zeolitic tuff packed columns operating under the same linear flow rates and concentration range.

Acknowledgments

Special thanks to late Dr. F. Mumpton for standard clinoptilolite samples. We thank Dr. Mehmet Polat, H. Demir, B. Cansever, G. Narin, Y. Akdeniz for their contribution in preparation of the representative clinoptilolite samples.

This study was supported by Turkish State Planning Organization (Project number 98K122130).

References

1. Inglezakis, V.J.; Loizidou M.D.; Grigoropoulou H.P. Equilibrium and kinetic ion exchange studies of Pb^{2+}, Cr^{3+}, Fe^{3+} and Cu^{2+} on natural clinoptilolite. *Water Research* **2001**, 00, 1-10.
2. Peric, J.; Trgo, M.; Medvidovic, N.V.; Removal of zinc, copper and lead by natural zeolite- a comparison of adsorption isotherms. *Water Research* **2004**, 38, 1893-1899.
3. Armbruster, T.; Simoncic, P.; Döbelin, N.; Malsy, A.; Yang, P.; Cu^{2+}-acetate and Cu^{2+}-amine exchanged heulandite. *Microporous and Mesoporous Materials* **2003**, 57, 121-131.
4. Argun, M.E.; Use of clinoptilolite for the removal of nickel ions from water: kinetics and thermodynamics. *Journal of Hazardous Materials* **2008**, 150, 587-595.
5. Kocasoy, G.; Sahin, V.; Heavy metal removal from industrial waste water by clinoptilolite. *Journal of Environmental Science and Health* **2007**, 42, 2139-2146.
6. Stylianou, M.A.; Inglezakis, V.J; Moustakas, K.G.; Malamis, S.P.; Lozidou, M.D. Removal of Cu(II) in fixed bed and batch reactors using natural zeolite and exfoliated vermiculate as adsorbents. *Desalination* **2007**, 215, 133-142.

7. Doula, M.; Ioannou, A.; Dimirkou, A.; Copper adsorption and Si, Al, Ca, Mg and Na release from clinoptilolite. *Journal of Colloid and Interface Science* **2002**, 245, 237-250.

8. Doula, M.K.; Ionnou, A.; The effect of electrolyte anion on Cu adsorption-desorption by clinoptilolite. *Microporous and Mesoporous Materials* **2003**, 58, 115-130.

9. Kocaoba, S.; Orhan, Y.; Akyuz, T. Kinetics and equilibrium studies of heavy metal ions removal by use of natural zeolite. *Desalination* **2007**, 214, 1-10.

10. Sprynskyy, M., Buszewski, B; Terzyk, A.P. Namiesnik, J. Study of the selection mechanism of heavy metal (Pb2+, Cu2+, Ni2+, and Cd2+) adsorption on clinoptilolite. *Journal of Colloid and Interface Science* **2006**, 304, 21-28.

11. Warchol, J.; Petrus, R. Modeling of heavy metal removal in packed beds, *Microporous and Mesoporous Materials* **2006**, 93, 29-39.

12. Stylianou, A.; Hadjiconstantinou, M.P.; Ingzelakis, V.J.; Moustakas, K.G.; Loizidou, M.D.; Use of natural clinoptilolite for the removal of lead, copper and zinc in fixed bed column. *Journal of Hazardous Materials* **2007**, 143, 575-581.

13. Chipera, S.J.; Bish, D.L.; A full-pattern quantitative analysis program for X-ray powder diffraction using measured and calculated patterns. *Applied Crystallography* **2002**, 35, 745-749.

14. Al-Jaroudi, S.S.; Ul-Hamid, A.; Mohammed, A.R.I.; Saner, S. Use of X-ray diffraction for quantitative analysis of carbonate rock reservoir samples. *Powder Technology* **2007**, 175, 115-121.

15. Cerri, G.; de' Gennaro, M.; Bonferoni, M.C.; Caramella, C.; Zeolites in biomedical application: Zn-exchanged clinoptilolite-rich rock as active carrier for antibiotics in anti-acne topical therapy. *Applied Clay Science* **2004**, 27, 141-150.

16. de Gennaro, B.; Colella, A.; Aprea, P.; Colella, C.; Evaluation of an intermediate-silica sedimentary chabazite as exchanger for potentially radioactive cations. *Micropororus and Mesoporous Materials* **2003**, 61, 159-165.

17. Chipera, S.J.; Bish, D.L. Multi-reflection RIR and Intensity Normalizations for Quantitative Analyses: Applications to Feldspars and Zeolites. *Powder Diffraction* **1995**, 10, 47-55.

18. Top, A.; Ülkü, S. Silver, Zinc, and Copper Exchange in a Na-Clinoptilolite and Resulting Effect on Antibacterial Activity. *Applied Clay Science* **2004**, 27, 13-19.

19. Duvarcı, O.C; Akdeniz, Y.; Özmıhçı, F.; Balköse, D.; Çiftçioğlu, M.; Ülkü S. Thermal Behaviour of a Zeolitic Tuff. *Ceramics International* **2007**, 33, 795-801.

20. Akdeniz, Y.; Ulku, S. Microwave effect on ion exchange and structure of clinoptilolite. *Journal of Porous Materials* **2007**, 14, 55-60.

21. Türkmen, M.; Removal of heavy metals from waste waters by use of natural zeolites. M.S. Thesis, İzmir Institute of Technology, **2001**.

22. Mondela, K.D.; Carland, R.M.; Aplan, F.F. The comparative ion exchange capacities of natural sedimentary and synthetic zeolites. *Minerals Engineering* **1995,** 8, 535-548.

23. Cincotti, A.; Lai, N.; Orru, R.; Cao, G. Sardinian natural clinoptilolites for heavy metals and ammonium removal: experimental and modeling. *Chemical Engineering Journal* **2001,** 84, 275-282.

24. Nikashina, V.A.; Tyurina, V.A.; Mironova, L.I. Sorption of copper (II) ions on the sodium and the calcium form of zeolites. *Journal of Chromatography* **1984,** 201, 107-112.

25. Guansheng, Z.; Xingzheng, L.; Guangju, L.; Quanchang, Z.; Removal of copper from electroplating effluents (potch water) using clinoptilolite, Occurence, properties and utilization of natural zeolites, *Budapest: Akademiai Kiado* **1988,** 529-539.

26. Langella, A.; Pansini, M.; Cappeletti, P.; de Gennaro, B.; de' Gennaro M.; Colella, C. NH_4^+, Cu^{2+}, Zn^{2+}, Cd^{2+} and Pb^{2+} exchange for Na^+ in a sedimentary clinoptilolite, North Sardinia, Italy. *Microporous and Mesoporous Materials* **2000,** 37, 337-343.

Chapter 7

Molecular Design of Thermally Responsive Metal Affinity Hydrogels for Contaminant Detection and Removal from Wastewater

Arunan Nadarajah, and Ganesh Iyer

Department of Chemical & Environmental Engineering, University of
Toledo, Toledo, OH 43606

Environmentally sensitive hydrogels have the ability to filter
large quantities of water in their swollen state and shrink to a
compact form in the collapsed state. Combining this property
with affinity groups offers the potential of a material with
excellent properties for waste water treatment. However, the
addition of other groups usually diminishes the
environmentally sensitive swelling properties. In this study a
method is developed to overcome these limitations employing
the monomer N-isopropyl acrylamide (NIPAAm) and a metal
affinity group. Employing a molecular design approach and
carefully balancing the hydrophobic and hydrophilic groups a
new co-monomer was synthesized and copolymerized with
NIPAAm to produce an affinity hydrogel. By controlling the
fraction of the co-monomer and the crosslinker it is shown that
this hydrogel can be made to mimic the thermally sensitive
phase transition behavior of pure NIPAAm hydrogels. The
gels also retain strong metal affinity with ~75% of the affinity
groups binding to copper ions.

Introduction

As described in many chapters in this volume, the increasing amounts and
variety of contaminants requires the development of newer approaches to their
removal. This is also necessitated by drawbacks of some of the current

contaminant removal technologies. Hydrogels with their ability to absorb many times their weight of water have found a variety of applications, such as drug delivery, bioseparations, chemical sensing (*1-5*). Among these is a relatively new class of hydrogels with environmentally sensitive swelling, where the swelling is triggered by changes in an external variable, such as temperature, salt concentration or pH.

The environmental sensitivity is an essential attribute to developing contaminant removal applications employing hydrogels. The swelling property of hydrogels enables them to filter large quantities of waste water and potentially harvest contaminants. However, the high degree of swelling causes the material to be extremely fragile, making its subsequent processing for contaminant removal a serious challenge. Unlike regular hydrogels, the environmentally sensitive ones can be collapsed into a more compact structure by changing an external variable. This allows the creation of a system that can efficiently filter wastewater and adsorb contaminants, easily remove the harvested contaminants in compact form, and then return the hydrogels for use again following contaminant release and regeneration.

Developing an environmentally sensitive hydrogel that can also harvest contaminants does pose some serious challenges. The origin of the environmental sensitivity is the careful balance between hydrophilic and hydrophobic groups in these hydrogels, which is easily upset by the addition of other groups. Their use in contaminant removal systems requires the addition of functional groups to bind the contaminants. Previous efforts to functionalize these hydrogels have caused the environmental sensitivity of these gels to diminish (*2,6*). An example of this is our recent study to develop environmentally sensitive hydrogels with metal affinity properties (*7*). The thermally responsive *N*-isopropyl acrylamide (NIPAAm) hydrogels were functionalized with metal affinity groups. While the resulting polymer did display both thermally responsive swelling and binding of metal ions, the swelling property was significantly compromised. Moreover, the functionalization of metal binding was found to be non-uniform, mostly occurring on the surface and not in the gel interior.

To overcome these challenges, a new approach is needed for the development of environmentally sensitive hydrogels for contaminant removal. In this study, we continue our focus on modified NIPAAm hydrogels. While thermally sensitive hydrogels may not be the best suited ones for contaminant removal, they provide an excellent model system for further development, as their properties are well understood. Similarly, different metal ions are a common contaminant, and metal affinity groups provide a well understood system for binding them. These have been widely used to provide metal affinity properties to polymers, including for wastewater treatment (*8,9*). The challenge here is to find a way to combine the metal affinity property with the thermally sensitive swelling property in a simple material without significantly compromising either one. If successful, this approach can be used to develop hydrogels with sensitivity to other variables functionalized with groups that can bind to other types of contaminants.

Molecular Design

The key to retaining environmental sensitivity in modified hydrogels is the retention of the hydrogel's balance of hydrophobic and hydrophilic groups (*10,11*). The hydrophilic groups are needed to attract water molecules into the pores within the gel, while the hydrophobic groups induce the gel to swell by causing the water molecules to form pentagonal clathrate groups around them. Retaining this behavior requires the design of a metal affinity group for inclusion in the gel with a similar hydrophoblic/hydrophilic balance as the NIPAAm monomer (*12,13*). In this case we designed a molecule with (a) a metal affinity iminodiacetic acid (IDA) group, (b) an acrylate group similar to NIPAAm for attachment to the polymer and (c) a hydrophobic hydrocarbon chain in the middle. We have called this molecule vinyl terminated iminodiacetic acid (or VIDA), and it is shown in Figure 1, along with the NIPAAm monomer (*14*). The hydrophobic hydrocarbon chain in VIDA balances the strongly hydrophilic IDA affinity group. This balance attempts to mimic that in the NIPAAm molecule. Another consideration in the molecular design is the effect of the metal binding itself on the hydrophobic/hydrophilic balance in the hydrogel. Divalent metal ions complex with IDA groups and neutralize them, thereby reducing the hydrophilicity of the hydrogel.

The other innovation in the development of the new hydrogel is in the functionalization process itself. To overcome the non-uniformity in the distribution of the affinity group in the hydrogel, the functionalization was carried out on an NIPAAm-like itself acrylate to produce the VIDA molecule as shown in Figure 1. By copolymerizing this molecule with NIPAAm, it is hoped that a hydrogel with sufficient metal affinity groups that are uniformly distributed in it will be produced.

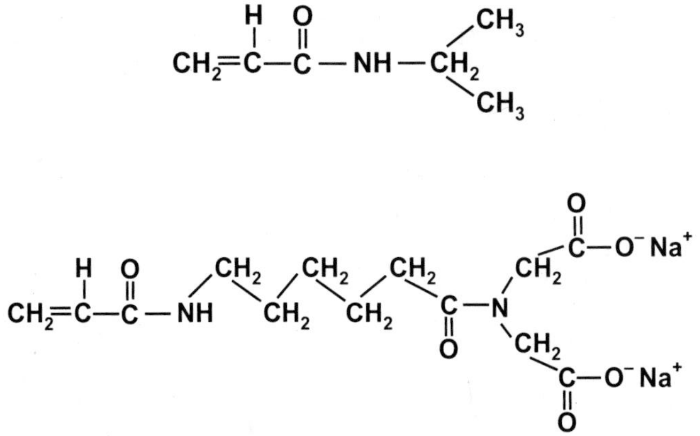

Figure 1. The monomers used in the synthesis of the VIDA-NIPAAm copolymer hydrogel. Top is NIPAAm and the bottom is the newly designed VIDA.

Material and Methods

Gel Synthesis and Functionalization

The thermally responsive monomer NIPAAm and crosslinker N,N–methylene–bisacrylamide (MBAAm, ultra pure grade) were purchased from Polysciences, Inc. The co-monomer N–(6–(acrylamido)hexanoyl)–iminodiaceticacid sodium salt (VIDA) was synthesized according to the procedure discussed in an earlier study (14). The photoinitiator riboflavin and accelerator N,N,N,N–tetramethylethylenediamine (TEMED) were obtained from Sigma–Aldrich. Copolymer hydrogels of NIPAAm and VIDA were synthesized with different weight percentages of comonomer VIDA and crosslinker MBAAm, while keeping the total weight of NIPAAm and VIDA as 1,500 mg. An example of this is the 15% (w/v) solution, prepared by dissolving 1300 mg of NIPAAm, 200 mg of VIDA and 55 mg of MBAAm in 10 ml of DI water, used to synthesize some of the gels. The solution was evacuated for 4 min to remove any dissolved oxygen following which 50 µl of 0.1% (w/v) riboflavin solution and 8 µl of TEMED were added. The reaction mixture was transferred between two dimethyldichlorosilane coated glass plates and irradiated with UV light for 2 hrs at room temperature (14).

The NIPAAm–VIDA hydrogels so obtained were cut into 1.6 cm discs and immersed in DI water for at least three days with frequent water changes to remove any unreacted monomers. Thoroughly dried samples were digested in nitric acid and hydrogen peroxide using standard protocol, in a microwave. To determine whether the VIDA monomer copolymerized with NIPAAm moieties during gel formation, the Na^+ content of the VIDA–NIPAAm gel was measured by induction coupled plasma analysis analysis (ICP, Thermolectron IRIS Interpid II), since it is the only component which can contribute sodium atoms to the gel. The gels were then incubated in 0.05M $CuSO_4$ solution for 24 hrs at room temperature to obtain copper chelated gels of Cu–VIDA–NIPAAm. The Cu–VIDA–NIPAAm gels were repeatedly washed with DI water for 3 to 4 days and then swollen and shrunk 2 times in DI water by increasing and decreasing the temperature between 5 and 50°C to ensure that there was no unbound copper entrapped in them. Copper contents in the gels were measured using ICP analysis.

Equilibrium Swelling Studies

The equilibrium swelling of the hydrogels in DI water was determined by recording the relative change in their diameter with temperature. The gel samples were equilibrated in 150-200 ml of the solution of interest for 24 hr at 20°C before the experiment. The change in the diameter of the gel between 20°C and 80°C was determined after equilibrating it at each intermediate temperature for at least 2.5 hrs. A temperature controlled Neslab GP300 water bath attached to a Neslab FTC 350 chiller was used to vary the temperature and a Gaertner Scientific X-axis measuring microscope was used to measure the diameter of the gel. Assuming isotropic behavior, the relative change in volume, V/V_0 can be calculated using equation $V/V_0 = (D/D_0)^3$, where, V_0 and D_0 are the equilibrium volume and diameter of the gel just after synthesis and V and D are the final volume and diameter of the gel. The onset temperature, which is defined as the point of intersection of the tangents to the curve at the beginning of the phase transition was used as the phase transition temperature also known as the lower critical solution temperature (LCST).

Results and Discussion

The VIDA-NIPAAm copolymer hydrogels had excellent properties visually when compared to pure NIPAAm hydrogels functionalized with metal affinity IDA groups (7). These hydrogels were highly transparent and when bound with Cu^{2+} ions displayed a uniform blue color over the entire hydrogel as shown in Figure 2. This suggests that the VIDA groups were indeed uniformly distributed in these gels, unlike the earlier efforts to functionalize NIPAAm hydrogels which resulted in significant non-uniformities in the distribution of affinity groups (7). This was confirmed by elemental analysis employing ICP as will be discussed later.

The most important consideration in this study was the swelling behavior of the VIDA-NIPAAM hydrogels. Figure 3 shows the swelling behavior for pure NIPAAm hydrogels as well as for VIDA-NIPAAm hydrogels with 2 different

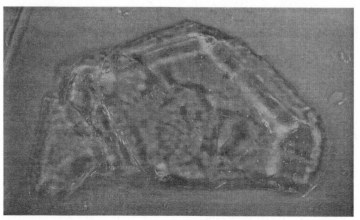

Figure 2. Freshly synthesized and equilibrated VIDA-NIPAAm copolymer hydrogel displaying a high degree of transparency and uniform distribution of copper ions.
(see page 2 of color insert)

amounts of VIDA. As mentioned before, the VIDA group is more hydrophilic than NIPAAm because of the two IDA groups (15,16). As the amount of VIDA is increased, it makes VIDA-NIPAAm as a whole more hydrophilic, shifting the phase transition temperatures higher (17). With 300 mg of VIDA, the enhanced hydrophilicity means that the hydrogel does not completely collapse even at elevated temperatures.

As mentioned earlier, the hydrophobic/hydrophilic balance is responsible for the temperature sensitive swelling behavior, with higher temperatures leading to a decrease of ordered water structures around the gel and the collapse of swelling. This means that in order to match the behavior of the pure NIPAAm hydrogels in Figure 3, the hydrophilicity of VIDA-NIPAAm gels must be reduced. One approach to achieve this is to increase the amount of hydrophobic crosslinker used during gel synthesis, which also has the added benefit of increasing its mechanical stability.

Figure 4 shows the effect of crosslinker amounts on the swelling behavior of the VIDA-NIPAAm hydrogels. While the crosslinker is a hydrophobic molecule, it is used sparingly and does not affect the hydrophobic/hydrophilic balance of VIDA-NIPAAm hydrogels much. However, it alters the rigidity of the hydrogels and its swelling behavior. In fact crosslinkers are a critical component in ensuring the mechanical stability of swollen hydrogels, and adequate amounts are needed for this. Crosslinkers can also reduce the porosity of the affinity hydrogel, thereby reducing its adsorption property, but this is not an issue here because of the relatively low amounts of crosslinker employed (14), and because the targets for adsorption are small metal ions.

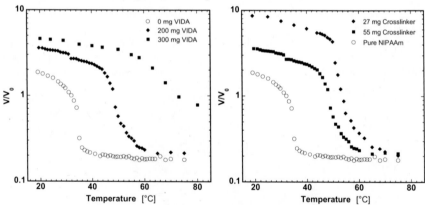

Figure 3. Swelling of pure NIPAAm and VIDA-NIPAAm copolymer hydrogels with 200 and 300 mg of VIDA as a function of temperature. The crosslinker amount used in the synthesis of the 3 gels is 55 mg.

Figure 4. The swelling of pure NIPAAm hydrogel, with 55 mg of crosslinker, and the VIDA-NIPAAm copolymer hydrogels, with 27 and 55 mg of crosslinker and 200 mg of VIDA, as a function temperature.

With higher crosslinker amounts, the gel swells less and the phase transition occurs at lower temperatures (18,19). This suggests that higher crosslinker concentrations can only partially balance out the high phase transition temperatures caused by the addition of VIDA monomers to the copolymer hydrogel. Additionally, Figure 4 shows that increasing the crosslinker amount gradually makes the hydrogel lose the sharpness of the phase transition, which will diminish the applicability of the gel for contaminant removal applications, so this has to be used judiciously.

The next variable to consider here is the binding of metal ions to the hydrogel itself. This is shown in Figure 5 where the swelling of VIDA-NIPAAm hydrogels with different VIDA concentrations when brought into contact with Cu^{2+} solutions is shown. When compared with the swelling of these hydrogels without the bound copper shown in Figure 3, it is clear that the metal binding changes the phase transition behavior significantly. The phase transition temperature decreases and approaches that of pure NIPAAm hydrogels. As mentioned earlier, this is caused by the Cu^{2+} complexing with the IDA groups in the copolymer hydrogel resulting in its decreased hydrophilicity. This suggests that our goal of producing a metal affinity hydrogel while retaining the phase transition behavior has indeed been achieved. In other words through careful balancing of hydrophilic and hydrophobic groups it is possible

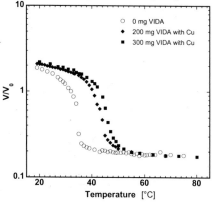

Figure 5. Swelling of pure NIPAAm and the VIDA-NIPAAm copolymer hydrogels shown in Figure 3 as a function temperature, but with bound copper ions.

to produce affinity hydrogels that fully retain their environmentally sensitive swelling properties.

The final remaining issue with these hydrogels is the extent of metal ion binding to them. To determine this, the VIDA-NIPAAm hydrogels were subjected to two stages of elemental analysis using ICP. In the first analysis, thoroughly dried samples of VIDA-NIPAAm were utilized to measure the amounts of Na^+ ions. The synthesis of VIDA-NIPAAm hydrogels produces sodium salts of the IDA groups allowing them to be quantified through an analysis of the bound Na^+ ions. By carrying out this analysis in different sections of hydrogel samples it was found that VIDA groups were uniformly distributed throughout the samples. Approximately 50% of the initial VIDA amounts undergo reaction during copolymerization.

Next, the amount of copper ions complexing with the swollen VIDA-NIPAAm hydrogels was analyzed. The hydrogels were fully hydrated and then equilibrated with $CuSO_4$ solution. To ensure that unbound Cu^{2+} were not included in the estimate, the gels were repeatedly swollen and shrunk times to delete the excess Cu^{2+}. The results showed that ~75% of the VIDA groups became chelated with Cu^{2+} ions, which is a high value for a polymer with metal affinity groups. Such a hydrogel can be highly effective in removing divalent cations from contaminated water (*14*).

These results indicate environmentally sensitive hydrogels have excellent properties making them suitable for use in devices for the purification of contaminated water. However, these properties have been demonstrated here only for thermally sensitive hydrogels with metal affinity groups. To make them truly versatile, this has to be extended to other environmentally sensitive gels, such as pH or light sensitive ones which may be more applicable for water purification. Similarly, other affinity groups need to be added to these hydrogels for the clean up of numerous organic and biological contaminants that are increasingly being found in waterways.

Conclusion

The phase transition behavior of thermally responsive hydrogels is highly sensitive to their hydrophobic/hydrophilic balance. Adding affinity groups to them using traditional functionalization techniques usually diminishes the phase transition behavior and also results in nonuniform distribution of these groups. These disadvantages have hindered the development of affinity hydrogels. For the case of NIPAAm based gels with metal affinity IDA groups considered here the alternate molecular design approach can produce hydrogels without these drawbacks. This is accomplished by developing the co-monomer VIDA with IDA groups with a similar hydrophobic/hydrophilic balance as NIPAAm, and copolymerizing with NIPAAm to produce the VIDA-NIPAAm copolymer hydrogel. By suitably varying the amounts of VIDA and crosslinker in the hydrogel it is possible to produce hydrogels with phase transition behavior that closely mimics those of pure NIPAAm hydrogels. In addition to retaining their phase transition behavior these hydrogels display strong metal affinity properties with ~75% of the affinity groups binding to copper ions in solution. By extending the techniques developed here to other environmentally sensitive gels and to other affinity groups it should be possible to develop versatile and effective systems for contaminant removal in waste water using these materials.

References

1. Zhuang, Y.; Chen, L.; Zhu, Z.; Yang, H. *Polym. Adv. Technol.* **2000,** *11*, 192.
2. Dong, C.L.; Hoffman, A.S. *J. Controlled Release* **1986,** *4*, 223.
3. Hoffman, A.S. *J. Controlled Release* **1987,** *6*, 297.
4. Kokufuta, E.; Aman, Y. *Polym. Gels Networks* **1997,** *5*, 439.
5. Harmon, M.E.; Tang, M.; Frank, C.W. *Polymer* **2003,** *44*, 4547.
6. Kaneko, Y.; Yoshida, R.; Sakai, K.; Sakurai, Y.; Okano, T. *J. Memb. Sci.* **1995,** *101*, 13.

7. Iyer, G.; Yoon, Y.S.; Coleman, M.R.; Nadarajah, A. *J. App. Poly. Sci.* **2007**, *105*, 1210.
8. Seko, N.; Tamada, M.; Yoshii, F. *Nucl. Instr. Meth. Phys. Res. B* **2005**, *236*, 21.
9. Wang, C.-C.; Chen, C.-Y.; Chang, C.-Y. *J. App. Poly. Sci.*, **2002**, *84*, 1353.
10. Pei, Y.; Chen, J.; Yang, L.; Shi, L.; Tao, Q.; Hui, B.; Li, J. *J. Biomater. Sci. Poly. Edn.* **2004**, *15*, 585.
11. Varghese, S.; Lele, A.K.; Mashelkar, R. A. *J. Chem. Phys.* **2000**, *112*, 3063.
12. Park, T.G.; Hoffman, A.S. *Macromolecules* **1993**, *26*, 5045.
13. Badiger, M.V.; Lele, A.K.; Bhalerao, V.S.; Varghese, S.; Mashelkar, R.A. *J. Chem. Phys.* **1998**, *109*, 1175.
14. Iyer, G.; Iyer, P.; Tillekeratne, L.M.V.; Coleman, M.R.; Nadarajah, A. *Macromolecules* **2007**, *40*, 5850.
15. Inomata, H.; Goto, S.; Saito, S. *Macromolecules* **1990**, *23*, 4887.
16. Işik, B. *J. App. Poly. Sci.* **2004**, *19*, 1289.
17. Liu, H.; Avoce, D.; Song, Z.; Zhu, X.X. *Macromol. Rapid Commu.* **2001**, *22*, 675.
18. Xisheng, Y.; Shuixin, T.; Yishi, S. *Chin. J. Poly. Sci.* **1990**, *8*, 224.
19. Kutsunori, T.; Toshikazu, T.; Toshiro, M. *J. Chem. Phys.* **2004**, *120*, 2972.

Chapter 8

Impact of Calcium on Struvite Precipitation from Anaerobically Digested Dairy Wastewater

Tianxi Zhang[1], Keith E. Bowers[2], Joseph H. Harrison[3] and Shulin Chen[1,*]

1. Department of Biological Systems Engineering, Washington State University, Pullman, WA 99164
2. Multiform Harvest Inc., Seattle, WA 98108
3. Department of Animal Sciences, Washington State University, Puyallup, WA 98371

The precipitation of struvite ($MgNH_4PO_4 \cdot 6H_2O$) from dairy wastewater is reviewed. There is a focus on conversion of the phosphorus (P) in calcium-phosphate solids into struvite, with emphasis on the P liberation into solution to be available as dissolved reactive phosphate ions prior to struvite formation. The P liberation was obtained using different methods such as acidification and sequestering calcium with a chelating agent. Struvite crystallization for P recovery in wastewater is discussed, with attention to both solution thermodynamics and kinetics, including struvite solubility, supersaturation degree, calcium inhibition, seed materials, and reactor types.

Struvite ($MgNH_4PO_4 \cdot 6H_2O$) formation is known in wastewater plants as a problem. Struvite deposits are present as blockages in pipes, centrifuges, belt presses, and heat exchangers and often cause system breakdowns. It is important to control and prevent struvite formation using various methods *(1)*, such as adding acids or chemical inhibitors. These methods add to the costs of wastewater treatment.

Although its formation is a problem in wastewater treatment, struvite could be recovered as a fertilizer because it is rich in phosphorus (P), nitrogen (N), and magnesium (Mg), thereby capturing a "waste" as a valued "product". Reducing P in wastewater has an important environmental implication. Excess P discharged from wastewater treatment can result in nutrient enrichment of surface water and result in algal blooms. Furthermore, P is a non-renewable resource as it is an important non-substitutable macronutrient in nature. Thus, struvite recovery would be a good option for P removal from wastewater.

An increasingly popular method of manure management on livestock farms is the use of anaerobic digestion (AD) technology with advantages as biogas production, odor reduction, pathogen removal, and stabilization of organic solids. However, the digested wastewater still has a high P concentration *(2)* because the AD process does not remove nutrients such as P. Thus, the digested wastewater seems to be a good source for P recovery. Struvite crystallization for P recovery has been investigated from animal wastewater, such as swine wastewater *(3, 4)*. Struvite crystallization can be influenced by several factors, such as solution pH, supersaturation of the three ions (Mg^{2+}, NH_4^+ and PO_4^{3-}) in the solution, and presence of impurities (e.g., calcium) *(5)*. It is noted that calcium in wastewater has a negative factor for struvite formation *(5)*. .

It is reported when calcium is present at high levels, the formation of struvite is inhibited because calcium-phosphorus precipitates are formed *(5, 6, 7)*. Unlike struvite, these calcium phosphate precipitates are not useful for reducing phosphorus content in the wastewater because their amorphous or near-amorphous character and very fine particle size render solids-removal techniques such as filtering and settling impractical. In addition, calcium phosphates have less value as fertilizer because they are nearly insoluble and because they contain no nitrogen or magnesium. There is limited information on the impact of calcium on struvite precipitation in wastewater, on the relationship between structure of calcium compounds and struvite formation, on the conversion methods of calcium compounds into struvite in wastewater, and on the influence of wastewater sources on struvite formation. The focus of this review is on struvite formation in the presence of high calcium content from anaerobically digested dairy wastewater.

1. Struvite Chemistry

Three soluble ions in equal molar concentrations, Mg^{2+}, NH_4^+ and PO_4^{3-}, react to form struvite according to the general chemical Equation 1 shown below.

$$Mg^{2+} + NH_4^+ + PO_4^{3-} + 6H_2O \rightleftharpoons MgNH_4PO_4 \cdot 6H_2O \qquad (1)$$

In general, struvite solubility is expressed as a solubility product, Ksp. A lower value of Ksp indicates lower concentrations of the three ions required for struvite formation. It is reported that the Ksp of struvite varies in the range of 5.50×10^{-14} to 3.98×10^{-10} *(8)*. Reported Ksp values are different for various complexes in solution and chemical speciation used by different authors (9). The typical value of Ksp was 2.51×10^{-13} (or pKsp of 12.6) used in literature, indicating low solubility of struvite. Thus, struvite formation requires low concentrations of the constituent ions, Mg^{2+}, NH_4^+, PO_4^{3-}.

High concentrations of the three constituent ions favor the rightward reaction in Equation 1, tending toward struvite formation. Supersaturation is used to measure the deviation of a dissolved salt from its equilibrium value. In order to represent supersaturation degree, the SI of a solution is defined as

$$SI = \log \left[\frac{IAP}{Ksp} \right] \qquad (2)$$

IAP is the product of the activities of the free constituent ions. SI values indicate tendency in the net direction of chemical reaction in Equation 1 because of the relationship between SI and the Gibbs free energy ΔG as given by Song et al., 2002 *(10)*:

$$\Delta G = - \left(\frac{2.303RT}{n} \right) SI \qquad (3)$$

where R is the ideal gas constant, T is the absolute temperature, and n is the number of ions in the precipitate compound. Based on chemical thermodynamic theory, the ΔG value can predict the reaction direction. Thus, SI is a good indicator of struvite precipitation based on the thermodynamic driving force.

It is common in wastewater to have ions such as calcium that can form compounds with P. In theory, an ion might have a negative influence on struvite precipitation if the compound reacts with any of Mg^{2+}, NH_4^+ or PO_4^{3-}, which decreases SI values. For example, a high level of calcium ions in wastewater would react with PO_4^{3-} ions to form calcium-phosphate compounds with low solubility, resulting in a low concentration of PO_4^{3-} ions. Low PO_4^{3-} ion concentration could result in the SI value below zero, and as a result, struvite precipitation would not occur. It has been reported that calcium at high levels in synthesized wastewater inhibited struvite formation (5).

Solution pH affects struvite solubility, largely by affecting the proportion of phosphate content present as free, non-protonated PO_4^{3-} ions. Generally, higher pH (e.g., pH 8.5) is favorable to struvite precipitation, as higher pH strips more protons from phosphate species. Thus, increasing solution pH is used as a method to enhance struvite formation *(11)*. However, a greater increase of pH (e.g. 10.0) might not favor struvite precipitation, since the ammonium ion concentration in solution would decrease as it also begins to undergo deprotonation to ammonia as pH approaches 9 or above. It was reported that the pH minimum of struvite solubility was 9.0-9.4 (12).

Another factor is temperature. Higher temperature increases struvite solubility (13). The lower temperature (e.g. 15 °C) favors struvite formation in compared to higher temperature (25 °C) (13). However, temperature above 64 °C could lead to structure changes of struvite crystals, resulting in decrease in struvite solubility *(14)*. The maximum of struvite solubility was at 50 °C *(14)*.

Although thermodynamic parameters described above, such as Ksp, supersaturation, are important, struvite crystallization is also associated with kinetic aspects including nucleation and growth.

2. Application of Struvite Crystallization in Wastewater

Both nucleation and growth are required for formation of struvite crystals. Nucleation is first phase, which occurs when ions combine to form crystal embryos. The second phase is growth, which is the process of constituent ions filling the crystal lattice of the embryos to form detectable crystals.

Rather than forming the crystal embryos, seeds are usually applied as nuclei for struvite crystallization in wastewater. Different types of seed materials have been investigated, including sand (15), struvite crystals (3), and steel mesh (16). Wang et al [2006] (17) compared three types of seeding materials, quartz sand, granite chips, and struvite crystals, in struvite crystallization using synthetic wastewater. Struvite crystals had the best performance on the phosphorus recovery among the three seeding materials tested (17). Physical properties of seeding materials, such as specific gravity and surface area, might be an important factor for struvite crystallization.

Struvite is useful in a size (e.g. 1.0 mm) of particles suitable for production in a crystallizer. Application of different types of crystallization reactors has been reported, including a mechanically stirring reactor (18), a gas agitated fluidized bed reactor (3), and water agitated fluidized bed reactor (19). Fluidized bed reactors could be suitable for wastewater, since suspended solids are present in the wastewater.

Struvite crystallization processes for P recovery have been extensively investigated in lab, pilot, and full scale models using synthesized and wastewater (4, 19, 20, 21). For example, Bowers and Westerman (2005) reported that total phosphorus (TP) removal (~80%) was obtained in field scale tests using the swine lagoon wastewater, a successful case for use of the struvite crystallization process for P recovery from animal wastewater.

However, struvite precipitation has not been proven effective in the presence of high calcium content, such as in anaerobically digested dairy wastewater. Recently, P recovery from flushed dairy manure wastewater was investigated in a fluidized-bed crystallization reactor (6). Products of calcium phosphate, not struvite, were obtained in the study, and confirmed by X-ray diffraction (XRD), scanning electron microscopy, and elemental analysis. The results suggested that the P was not available in the phosphate ionic form required for struvite formation since high level of calcium was present in the dairy wastewater.

3. Liberation of Phosphorus for Struvite Formation in Dairy Wastewater

Calcium does inhibit struvite formation if enough is present to tie up phosphate as calcium-phosphate solids in the wastewater. High levels of calcium are indeed present in anaerobically digested dairy wastewater (22). Before discussing how to convert calcium-phosphate particles into struvite, we need to know P chemistry, including P speciation and phase distribution in the dairy wastewater.

3.1 Phosphorus chemistry in dairy manure

High calcium levels in dairy manure are a result of the cow's diet. Table 1 presents the composition of diets in the different dairy operations (22). High calcium levels at 0.66-1.00 % were reported in the diets. A molar ratio of Ca : P was found as 1.66-2.43. Calcium, phosphorus and magnesium are all found in dairy manure at significant concentrations. Different types of Ca-P compounds could be present in animal manure. P in manure has been characterized and classified into inorganic, organic, phospholipid and extracted fractions *(23, 24, 25)*.

Table 1. Composition of the dairy cow diets

Name [a]	Ca (%)	Mg (%)	P (%)	Ca : P [b]	Mg : P [b]
DS	0.71-1.23	0.23-0.35	0.30-0.42	1.83-2.43	0.99-1.33
S	1.00	0.35	0.40	1.94	1.13
T	0.92	0.40	0.43	1.66	1.20
W1	0.86	0.36	0.36	1.85	1.29
W1	0.66-0.91	0.36-0.46	0.28-0.36	1.83-2.28	1.67-2.12

a abbreviations of different dairies

b molar ratio

SOURCE: Reproduced with permission from reference 22. Copyright.

It is important to know the P speciation and phase distribution in dairy manure in terms of ratio of inorganic P and organic P and dissolved versus particulate form as these fractions affect struvite formation.

Inorganic P in manure was the majority of the TP *(26)*, and the inorganic P was the predominant particulate-bound form *(27)*. The high calcium content in dairy manure contributes to particulate P with low solubility. It was strongly supported that inorganic P in animal manure is predominantly in Ca-P and/or Mg-P bound form from P extraction studies *(28, 29, 30)*. In addition, Güngör and Karthikeyan (2008) (22) have also evaluated influence of AD process of dairy manure on P speciation and phase distribution. Their results further confirmed that the majority of TP in dairy manure was in a particulate bound form. Only about 7% of TP was in a dissolved form in the digested effluent while total dissolved P (TDP) constituted about 12% of TP in the undigested influent. The AD process could mineralize dissolved unreactive P (DUP) and subsequently partition dissolved phosphate ions into a particle-bound form, resulting in a slight increase in the particulate P content of digested manure.

The specific P speciation present in the dairy manure wastewater is not clear. Geochemical equilibrium modeling with software Mineql+ was used to determine the probable P speciation. Recently, Güngör and Karthikeyan (2008) (22) suggested that dicalcium phosphate dehydrate, dicalcium phosphate anhydrous, octacalcium phosphate, newberyite, and struvite were probable solid phases in both the digester influent and effluent from the Mineql+ simulation. Their results indicated that AD did not influence the solid P type and stability.

128

It should be noted that different dairy operations could affect manure characteristics (such as ionic composition), further affecting the distribution of the P solid phases (22).

Nevertheless, the majority of the P in digested dairy wastewater is in a suspended solid form because of the high Ca levels. The P in the Ca-P particulate solids cannot directly be converted into struvite because struvite formation requires dissolved phosphate ions presented in Equation 1. One of the approaches for this was to liberate the P from the Ca-P particulate solids and subsequently form struvite precipitates.

3.2 Liberation of phosphorus from anaerobically digested dairy wastewater

Several methods were studied to release the P in the Ca-P solids into dissolved phosphate ions, such as acidification and addition of a chelating agent (7).

3.2.1 Acidification

Acidification was used to dissolve the P into solution as soluble phosphate ions by protonating the phosphate ions. Solid calcium phosphate could be dissolved when the calcium-phosphate ionic product is below the equilibrium solubility product for calcium phosphates. Inorganic acid, such as concentrated HCl, was directly added to the wastewater samples to depress the pH to various targeted degrees [Zhang et al., 2008].

The results of the acidification to dissolve P and Ca^{2+} ions in solution are shown in Figure 1 (7). From this figure, concentrations of Ca^{2+}, TDP, and DRP increased significantly with pH decrease. For example, the concentrations of Ca^{2+} and DRP of 1.1 and 0.5 mmol/L at original pH 7.8 were increased to 17 and 7.4 at pH 3.8, respectively. These results demonstrated that more P was solubilized into solution as pH decreased. Thus acidification released P and Ca^{2+} into the solution.

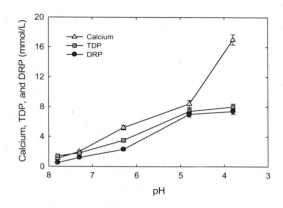

Figure 1. Concentrations of calcium, TDP, and DRP in the solution after acidification (Reproduced with permission from reference 7. Copyright)

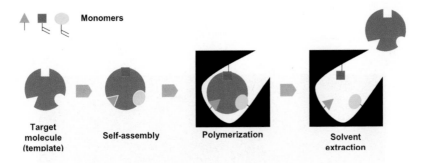

Figure 3.1. Schematic representation of the non-covalent approach for molecular imprinting.

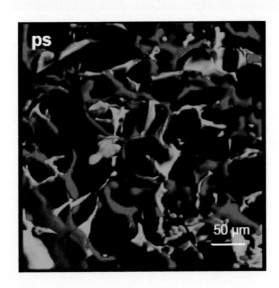

Figure 3.3. Confocal Laser Scanning Microscopy (CLSM) 3D reconstituted images of macroporous cryogel with double-continuous macroporous networks and stained with fluorescent dyes. The primary PEG-cryogel was stained with Rhodamine B (red colored walls) and the secondary PEG-cryogel was stained with FITC (green colored walls). Reproduced from (47) with permission.

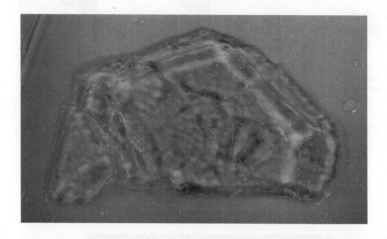

Figure 7.2. Freshly synthesized and equilibrated VIDA-NIPAAm copolymer hydrogel displaying a high degree of transparency and uniform distribution of copper ions.

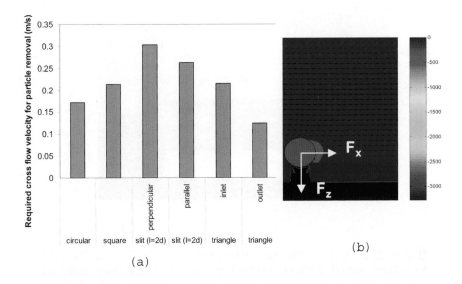

(a)

(b)

Figure 9.2. (a) Required cross flow velocity for particle removal from various pore geometries. (b) Simulation box for particle removal from a pore.

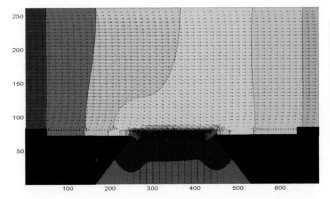

Effect of channel height on flux	
Height (micron)	Flux (%)
35	30
60	63
150	92
350	99

Figure 9.4. Effect of channel height on fluxes of microsieves with different substructures (colours represent different pressure ranges – red the highest, blue the lowest pressures).

Figure 11.5. Several different antifouling paints prepared from PAMAMOS dendrimers and HB-PUSOX. Note that because the base polymer solutions are colorless, different colors originate from different antifoulants used. Small vials on the tops of the respective paint containers contain catalyst solutions that may or may not be needed (see text for explanation).

The released phosphate ions might not form struvite when the solution pH rises to the original pH of 7.8. These phosphate ions in a low pH solution could return back to calcium-phosphate particles, as concentrations of the constituent ions in the probable Ca-P species are likely to exceed their equilibrium solubility at pH of 7.8. For example, Ca-P solids would be formed if the calcium content is so high that the equilibrium moves toward the Ca-P compounds. In a similar way, struvite could be formed if constituent ions of struvite (e.g. NH_4^+ and Mg^{2+}) have high concentrations in the solution.

In addition, struvite formation is influenced by several other factors, such as thermodynamics of liquid-solid equilibrium, phenomena of mass transfer between solid and liquid phases, kinetics of reaction, solution pH, supersaturation, mixing energy, temperature, and presence of impurity ions (5). Calcium as an impurity ion was a negative factor on struvite crystal size, shape and purity by inhibiting the growth rate of struvite crystals due to the blocking of active growth sites by adsorption on the crystal surfaces (5). For example, an amorphous calcium phosphate, not a crystalline struvite, was formed at a molar ratio of Mg : Ca ~ 1 : 1 and above (5).

3.2.2 EDTA treatment

Another method for P liberation is to add a chelating agent into wastewater, such as ethylenediaminetetraacetic acid (EDTA). Calcium in the solid Ca-P compounds could be sequestered because EDTA has a stronger binding to calcium, with a high stability constant to $[EDTA-Ca]^{2-}$ of $10^{+10.7}$. Interaction between EDTA and Ca-P compounds is represented in Equation 4. EDTA added into the solution would react with the calcium when calcium is present in the solid form of $[Ca-PO_4]^-$. The reaction product is a complex of $[EDTA-Ca]^{2-}$ and PO_4^{3-} ions, both of which would dissolve in the solution as soluble ions. Thus, EDTA addition results in an increase of calcium concentrations in the $[EDTA-Ca]^{2-}$ form and P as phosphate ions in the solution.

$$[Ca-PO_4]^- \text{ (s)} + [EDTA]^{4-} \text{ (aq)} \rightleftharpoons [EDTA-Ca]^{2-} \text{ (aq)} + PO_4^{3-} \text{ (aq)} \qquad (4)$$

(s and aq represent solid and aqueous solution, respectively)

Figure 2 presents the EDTA results in liberating P into solution (7). As shown in Figure 2, more calcium and P could be dissolved in the solution with an increase in the amount of EDTA addition. For instance, most of the TP (91%) and calcium (93%) can be released into the solution by adding EDTA. It is noted that the release profile of magnesium lagged, but was similar to, that of calcium. This lag can be explained in that the stability of the $[EDTA-Ca]^{2-}$ complex is two orders of magnitude greater than that of the $[EDTA-Mg]^{2-}$ complex. Thus, EDTA starts to bind magnesium when almost all of the calcium is bound. The results demonstrated that the P that had been calcium-bound in the solids was released into solution *(7)*.

130

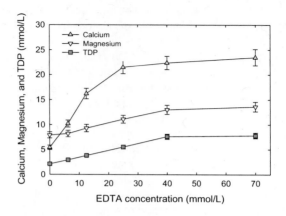

Figure 2. Concentrations of calcium, magnesium and TDP in the solution after EDTA addition (Reproduced with permission from reference 7. Copyright)

3.3 Struvite formation using anaerobically digested dairy wastewater

Although the released phosphate ions are available for struvite formation, it also requires free Mg^{2+} ions as they are the constituent ions of struvite. However, reduction of free Mg^{2+} ions occurred in high concentration of EDTA, as EDTA binds with magnesium as an $[EDTA-Mg]^{2-}$ complex with stability constants $10^{+8.7}$. Then, magnesium might be unable to form struvite precipitates at high EDTA levels.

To provide more free Mg^{2+} ions required by struvite formation, magnesium chloride was added to increase the magnesium ion molarity at different levels (7). At sufficient levels of Mg^{2+} ions, all EDTA would be bound and then excess Mg^{2+} ions favors struvite precipitation with available PO_4^{3-} ions and NH_4^+ ions present in solution. Zhang et al. (2008) (7) reported that the excess Mg^{2+} ions combined with the P to form the precipitates, resulting in a sharp reduction of P concentration in the solution.

After the precipitates were obtained, it was confirmed that the product was struvite. Solid elemental composition and X-Ray Diffraction (XRD) analyses are typically used to determine the struvite structure. Considering that theoretical molar ratio of P : N : Mg is 1 : 1 : 1, the solid composition of a molar ratio of P : N : Mg should be close to the theoretical ratio, such as the 1.0 : 1.06 : 1.05 reported in *(4)*. XRD is another method for determination of struvite structure, which depends on both diffraction angles and intensities. Zhang et al. (2008) *(7)* compared the XRD result of the precipitate sample obtained from the released P with that of standard struvite shown in Figure 3. The XRD pattern from the precipitated sample matched with the unique pattern of standard struvite very well on both diffraction angles (2θ) and intensities. Thus, the solid obtained was indentified as struvite based on the XRD pattern and the elemental composition *(7)*.

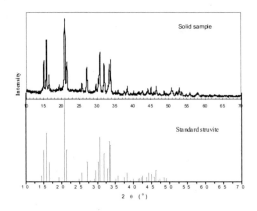

Figure 3. X-Ray Diffraction pattern from the solid product compared with standard struvite (Reproduced with permission from reference 7. Copyright)

3.4. Effects of Wastewater Characteristics on Struvite Crystallization

Wastewater characteristics affect performance of struvite crystallization. Three different wastewaters were compared for struvite crystallization using a cone-shaped fluidized bed struvite crystallizer at pilot scale *(31)*. The three wastewaters tested were (1) swine lagoon wastewater; (2) dairy lagoon wastewater; and (3) anaerobically digested dairy wastewater.

Characteristics on total solids (TS), calcium, magnesium and TP varied among the three different wastewaters. The swine lagoon wastewater contained about 0.2 % of TS, about 200 mg/L of calcium, 20-70 mg/L of magnesium, and 50-120 mg/L of TP. The dairy lagoon wastewater had about 0.4% of TS, 300 mg/L of calcium, about 100 mg/L of magnesium, and same range of TP as the swine wastewater. The dairy digested wastewater contained greater amounts of solids (about 3% of TS), calcium (about 950 mg/L), magnesium (about 330 mg/L), and TP (about 280 mg/L).

The characteristics of the three different wastewaters had a significant influence on struvite crystallization *(31)*. Greater than 70% removal of TP was achieved using the crystallizer system in the swine lagoon wastewater. For the dairy lagoon wastewater, phosphorus removal at first was near zero. Calcium in the Ca-P solids might have contributed to this poor performance. Thermodynamic estimation using these conditions indicated that, the dairy lagoon water contained sufficient Mg, Ca, ammonium, and P to supersaturate it with respect to struvite and/or calcium phosphate compounds such as beta-tricalcium phosphate (BTCP). To release P and make it available for struvite crystallization, acid was blended into the lagoon water in the surge tank to lower its pH to 5.0 as a pretreatment step. Then, the TP removal increased to 50 % in the dairy lagoon wastewater. The digested wastewater tested was higher in solids and nutrient content (e.g. calcium) than the swine and dairy lagoon water. However, TP removal rarely exceeded 25% using the same acid treatment as was used in the dairy lagoon wastewater. EDTA treatment was not used in the pilot tests described in *(31)* for the digester wastewater; however, at the symposium for which *(31) was prepared*, the authors presented data showing subsequent use of EDTA treatment increased TP removal to 70% or greater. The results of the three different wastewaters confirmed that it requires liberation of P into the dissolved phosphate ion prior to struvite crystallization in dairy wastewater.

4. Conclusion

Phosphorus in dairy manure wastewater was present as particulate calcium phosphate solids. A majority of phosphorus can be released into solution as phosphate ions by acidification and adding an EDTA chelating agent. Then, struvite was formed by the liberated P with available magnesium ions and ammonium ions. Both thermodynamic (e.g. supernatant degree) and kinetic (e.g. seed materials) properties of the wastewater also affected struvite crystallization. Work so far has demonstrated struvite precipitation in dairy wastewater at the lab and pilot scale. More research on P liberation and further struvite crystallization at full scale is necessary in order to evaluate the economical feasibility for P recovery from dairy wastewater.

Acknowledgements

This work was supported by a Conservation Innovation Grant from the U.S. Department of Agriculture Natural Resource Conservation Service (NRCS 68-3A75-4-201, 69-3A75-7-110).

References

1. Doyle, J. D.; Parsons, S.A. Struvite formation and control. *Water Sci. Technol.* **2004.** 49, 177-182.
2. Zhang, R. H.; Tao, J.; Dugba, P. N. Evaluation of two-stage anaerobic sequencing batch reactor systems for animal wastewater treatment. *Trans. ASAE* **2000,** 43, 1795-1801.
3. Bowers, K. E.; Westerman P. W. Performance of cone-shaped fluidized bed struvite crystallizer in removing phosphorus from wastewater. *Trans. ASAE* **2005,** 48, 1227-1234.
4. Suzuki, K.; Tanaka, Y.; Kuroda, K.; Hanajima, D.; Fukumoto, Y. Recovery of phosphorus from swine wastewater through crystallization. *Bioresour. Technol.* **2005,** 96, 1544-1550.
5. Le Corre, K. S.; Valsami-Jones, E.; Hobbs, P.; Parson, S. A. Impact of calcium on struvite crystal size, shape and purity. *J. Cryst. Growth* **2005,** 283, 514-522.

6. Harris, W. G.; Wilkie, A. C.; Cao, X.; Sirengo, R. Bench-scale recovery of phosphorus from flushed dairy manure wastewater. *Bioresour. Technol.* **2008**, 99, 3036-3043.

7. Zhang, T; Bowers, K. E.; Harrison, J. H.; Chen, S. Releasing Phosphorus from Calcium for Struvite Fertilizer Production from Anaerobically Digested Dairy Effluent. *Water Environ. Res.* **2008**. (Submitted).

8. Doyle, J. D.; Parson, S. A. Struvite formation, control and recovery. *Water. Res.* **2002**, 36, 3925-3940.

9. Buchanan, J. R.; Mote, C. R.; Robinson, R. B. Thermodynamics of struvite formation. *Trans. Am. Soc. Agric. Eng.* **1994**, 37, 1301-1308.

10. Song, Y., Hahn, H. H.; Hoffmann, E. Effects of solution conditions on the precipitation of phosphate for recovery: a thermodynamic evaluation. Chemosphere **2002**, 48, 1029-1034.

11. Battistoni, P.; Fava, G.; Pavan, P.; Musacco, A.; Cecchi, F. Phosphate removal in anaerobic liquors by struvite crystallization without addition of chemical: preliminary results. *Water Res.* **1997**, 31, 2925-2929.

12. Booker, N. A.; Priestley, A. J.; Fraser, I. H. Struvite formation in wastewater treatment plants, opportunities for nutrient recovery. *Environ. Technol.* **1999**, 20, 777-782.

13. Adnan, A.; Dastur, M.; Maviniv, D. S.; Koch, F. Preliminary investigation into factors affecting controlledd struvite crystallization at the bench scale. *J. Environ. Eng. Sci.* **2004**, 3, 195-202.

14. Aage, H. K.; Anderson, B. L.; Blom, A.; Jensen, I. The solubility of struvite. *J. Radioanal. Nucl. Chem.* **1997**, 223, 213-215.

15. Battistoni, P.; De Angelis, A.; Prisciandaro, M.; Boccadoro, R., Bolzonella, D. P removal from anaerobic supernatants by struvite crystallization: long term validation and process modeling. *Water Res.* **2002**, 36, 1927-1938.

16. Le Corre, K. S.; Valsami-Jones, E.; Hobbs, P.; Jefferson, B.; Parson, S. A. Struvite crystallization and recovery using a stainless steel structure as a seed material. *Water Res.* **2007**, 41, 2449-2456.

17. Wang, J.; Burken, J. G.; Zhang, X. Effect of seeding materials and mixing strength on struvite precipitation. *Water Environ. Res.* **2006**, 78, 125-132.

18. Pastor, L.; Mangin, D.; Barat, R.; Seco A. A pilot-scale study of struvite precipitation in a stirred tank reactor: Conditions influencing the process. *Bioresour. Technol.* **2008**, 99, 6285-6291.

19. Adnan, A.; Mavinic, D. S.; Koch, F. A. Pilot-scale study of phosphorus recovery through struvite crystallization – examining the process feasibility. *J. Environ. Eng. Sci.* **2003**, 2, 315-324.

20. Burns, R. T.; Moody, L. B.; Walker, F. R.; Raman, D. R. Laboratory and in situ reductions of soluble phosphorus in swine waste slurries. *Environ. Technol.* **2001**, 22, 1273-1278.

21. Jeong, Y. K.; Hwang, S. J. Optimum doses of Mg and P salts for precipitating ammonia into struvite crystals in aerobic composting. *Bioresour. Technol.* **2005**, 96, 1-6.

22. Güngör, K.; Karthikeyan, K. G. Phosphorus forms and extractability in dairy manure: a case study for Wisconsin on-farm anaerobic digesters. *Bioresour. Technol.* **2008**, 99, 425-436.

134

23. Barnett, G. M. Manure P fractionation. *Bioresour. Technol.* **1994,** 49, 149-155.
24. He, Z.; Honeycutt, C. W. Enzymatic hydrolysis of organic phosphorus in animal manure. *J. Environ. Qual.* **2001,** 30, 1685-1692.
25. Dou, Z.; Knowlton, K. F.; Kohn, R. A.; Wu, Z.; Satter, L. D.; Zhang, G.; Toth, J. D.; Ferguson, J. D. Phosphorus characteristics of dairy feces affected by diets. *J. Environ. Qual.* **2002,** 31, 2058-2065.
26. Sharpley, A. N.; Moyer, B. Phosphorus forms in manure and compost and their release during simulated rainfall. *J. Environ. Qual.* **2000,** 29, 1462-1469.
27. Gerritse, R. G.; Vriesema, R. Phosphate distribution in animal waste slurries. *J. Agric. Sci.* **1984,** 102, 159-161.
28. Chapuis-Lardy, L.; Temminghoff, E. J. M.; Goede, R. G. M. De. Effects of different treatments of cattle slurry manure on water-extractable phosphorus. *Netherlands J. Agric. Sci.* **2003,** 51(1-2), 91-102.
29. Güngör, K.; Karthikeyan, K. G. Influence of anaerobic digestion on dairy manure phosphorus extractability. *Trans. ASAE* **2005a,** 48, 1497-1507.
30. Güngör, K.; Karthikeyan K. G. Probable phosphorus solid phase and their stability in ananerobically digested dairy manure. *Trans. ASAE* **2005b,** 48, 1509-1520.
31. Bowers, K. E.; Zhang, T., Harrison, J. H. Phosphorus removal by struvite crystallization in various livestock wastewaters. *International Symposium on Air Quality and Waste Management for Agriculture.* September 15-19, **2007,** Broomfield, Colorado, USA.

Part II:
Advanced Membranes

Chapter 9

Design of Microsieves and Microsieve Processes for Suspension Fractionation

A.M.C. van Dinther, C.G.P.H. Schroën, R.M. Boom

Wageningen University, Department of Agrotechnology and Food Sciences, Food and Bioprocess Engineering group, Bomenweg 2, 6703 HD Wageningen, The Netherlands.

Micofiltration is a mature process for separating particles from a solution. Fractionation, that is the separation of larger from smaller particles, is however more difficult, especially when the components to separate are not very different in size (*1*). In this chapter we evaluate the use of micro-engineered membranes for fractionation. These membranes or microsieves are a new type of membrane of which the pore size and geometery can be tailored to an extend that is unprecedented in 'regular' membranes. In the design of a fractionation process with microsieves necessarily we consider a wide range of size scales, from the very small dimension of one pore all the way up to module-scale fluid dynamics. In the conclusion section, the different scales are compared, and their importance for the actual fractionation process evaluated.

Introduction

Microsieves: General Characteristics

Microsieves (Figure 1) are a novel type of membrane, manufactured from silicon wafers with photolithographic techniques (*2*). Because of their extremely thin active top layer, relatively high porosity, and open support structure, their clean-water fluxes can be two or three orders of magnitude higher than fluxes of conventional membranes, even at very low transmembrane pressures. The

138

topsurface is much smoother than for any other membrane, which may have advantages in relation to fouling and cleaning. Furthermore, photolithography allows great freedom in design of the pores (e.g. size, geometry), as well as positioning of the pores. Typical pore sizes are between 0.1 and 10 micron.

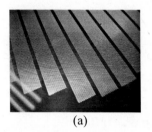 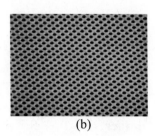 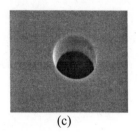

(a)   (b)   (c)

Figure 1. Microsieves from different scale perspective. (a) Part of a wafer with sieve fields and solid lanes, (b) porous field, (c) close-up of one pore. Pictures are courtesy of Aquamarijn Microfiltration BV, The Netherlands.

Given the uniform pore size and the possiblitiy to tune the system to the separation issue at hand, microsieves are very interesting for fractionation purposes. However, microsieve technology is still young and compared to conventional membranes, silicon-based microsieves are costly. The newest development, to make microsieves more price attractive, is in the field of polymeric microsieves (*3*). For now, we will only focus on the unique properties of the microsieves, and their additional value in fractionation.

In this chapter, practical and simulation results on silicon and polymer microsieves, as well as regular membranes will be presented. Mostly model systems will be reported upon, but in some cases also practical systems such as fractionation of dairy ingredients will be discussed. The first part of this chapter is dedicated to the design of the microsieve, and the second part will focus on effects of fluid dynamics.

Effects related to Microsieve Design

Effect of Pore Geometry on Particle Release

During microfiltration, particles are dragged with the flow towards the membrane, and may be captured on and in pores, thereby decreasing the flux. Inn order to maintain a reasonable flux, they need to be removed again. In addition to process parameters, the geometry of the membranes (3D pore shape, surface porosity) influences the force that keeps the particles to the membrane. Microsieves are ideal for investigating the influence of the membrane geometry, and one has large freedom in the design. A first investigation of the influence of geometry was done through computational fluid dynamics (more information see (*4*)). A particle is placed on a pore of a specific 3D shape and the (cross-flow) force which is needed to release the particle from the pore is calculated. A typical result for particles of 1 μm placed on pores of 1.2 μm, is shown in Figure 2a, together with a visualization of the system, including pressure distribution (colours) and velocities (arrows) (Figure 2b); the applied transmembrane pressure is 3.33 kPa (*4*).

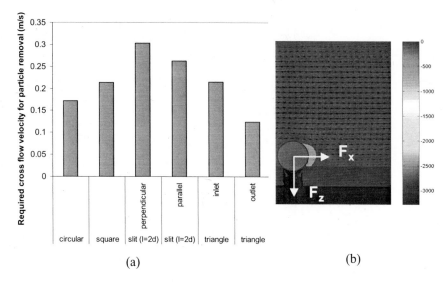

(a) (b)

Figure 2. (a) Required cross flow velocity for particle removal from various pore geometries. (b) Simulation box for particle removal from a pore.

(see page 3 of color insert)

The suction force (F_z) determines the shear force (F_x) needed to release the particle and it is clear that there is a big difference between the different geometries. For a rectangular pore, the suction force of the liquid which passes through that part of the pore which is uncovered by the particle, is relatively large and as a consequence a higher cross-flow velocity is needed to remove the particle. Surprisingly enough, a triangular pore shape with one angle pointing to the outlet, seems to allow easiest particle removal. In this case, most of the pore is covered by the particle, so the suction force is not very high. At the same time, the pivot point of the particle is no longer in the centre of the pore; it is considerably shorter, and therewith removal is facilitated.

Also the effect of the top surface of the microsieve was investigated. The sharper the edges of the pore, the easier the removal of particles is: with sharp edges the particles are farthest exposed to the cross-flowing liquid. If particles can settle deep inside the pore, the required cross-flow velocity increases considerably, since the pivot point is then deeper in the membrane and hence the arm for lifting is smaller. In an actual process, very sharp edges may in fact enhance adsorption of particles due to the high surface energy of very sharp edges. Therefore we expect that in real membranes, there will be a balance. Even though sharper pore edges may be advantageous, one migh expect that the edges will wear off from operation and cleaning, and become more rounded in

140

time. The process will therefore need faster cross-flowing to guarantee particle removal (*4*).

Effect of Multiple Particles on Particles Release

Obviously, in a microsieve many pores are present and many particles can block these pores. As soon as adjacent (upstream) pores are blocked, these blocking particles will then screen the flow around the next pores, and therefore the force exerted by the crossflow at these pores will be much smaller. This effect was investigated through computer simulations. The conditions are the same as for Figure 2, the pore geometry is circular. The effect of surface coverage is shown in Figure 3 (*4*).

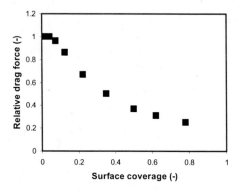

Figure 3. Effect surface coverage on drag force experienced by cross flow. Reproduced with permission from reference 4. Copyright 2006.

In Figure 3 it is clearly indicated that once some particles are captured on a membrane they become increasingly difficult to remove, due to mutual screening from the flow. On the x-axis the surface coverage is shown, defined as the cross-sectional area of spherical particles divided by the membrane area. The relative drag force on a particle can be seen on the y-axis. At 80% coverage, the drag-force that is experienced by the particles is 4 times less than if only a few particles were present, and this implies that the cross-flow should be increased accordingly. This is valid not only for removal of particles. Right behind a particle the next one will settle since it is shielded from the cross flow. This effect was also experimentally observed (*4*). Therefore in practice, one might expect a hysteresis effect. As long as blockage is very low, the crossflow may be sufficient to avoid severe further deposition, but as soon as a few particles are on the membrane surface, others will settle very quickly. One then needs a considerably higher crossflow velocity to remove the particles again.

Design of Microsieves

The microsieves have a specific design which matches high permeability with sufficient mechanical strength. Therefore, very thin sieve fields are surrounded by supporting non-porous areas (see Figure 1a), and substructures which are below the actual sieve. A schematic representation is given in Figure 4. The black part corresponds to the support structure, while the porous part of the sieve is shown as a dotted line which crosses the structure from the top of the black substructure.

Once more, the flux through a microsieve was investigated through CFD simulation, and a typical result is shown in Figure 4 (5). The cross-flow enters the channel on the left side of the picture accompanied by a high pressure (shown as red). Right below the microsieve, the pressure is practically the same as in the cross-flowing liquid, with the exception of the area just above the central permeate collection area (darkest blue area). In this area a bigger pressure drop exists, and most of the permeate is collected here. Since the central permeate collection area can be designed differently, we investigated what the channel height should be in order to use the membrane more efficiently. It turned out that the original height of 35 micron should be increased by a factor of 5 to 10 in order to reach at least 90% of the maximum flux and make best use of the available surface area. This shows that due to the very high permeabilities of the porous parts of the sieve, the design of the supporting substructure has strong influence on the performance of the total membrane.

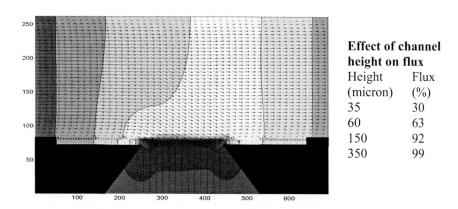

Effect of channel height on flux

Height (micron)	Flux (%)
35	30
60	63
150	92
350	99

Figure 4. Effect of channel height on fluxes of microsieves with different substructures (colours represent different pressure ranges – red the highest, blue the lowest pressures). (see page 3 of color insert)

142

Concentration Polarization and Cake Layer Formation Effects

In addition to the effects discussed above, larger scale effects are just as important. For this, the development of the concentration of particles above a single sieve field was simulated (for detailed information consult (6)). Concentration polarization and cake layer formation, built up by particles that cannot pass the pores, are both shown.

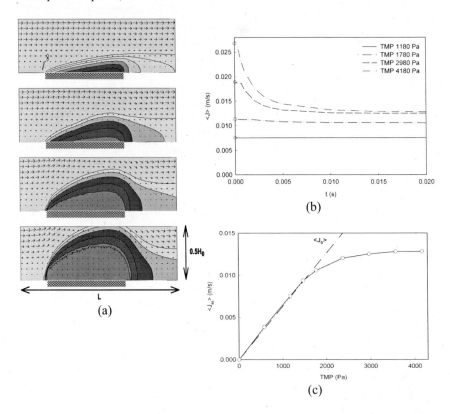

Figure 5. Illustration with CFD simulation of effects of transmembrane pressure on concentration polarization layer, with increasing transmembrane pressure (a), flux as a function of time and transmembrane pressure (b), steady state flux as function of transmembrane pressure (c). Reproduced with permission from reference 6. Copyright 2005.

The effect of the transmembrane pressure on concentration polarisation was investigated with a fixed crossflow velocity of 0.32 m/s; typical examples of simulation results are shown in Figure 5a. The channel height shown in the figure is only half of the total channel height. Here, transport of particles to the membrane is too low to cause cake layer formation. In Figure 5b and c, cake layer formation can be seen by flux decline as a function of pressure. In Figure 5b, the fluxes corresponding to the various pressures are given as a function of time and in Figure 5c the steady state flux is given as a function of the applied pressure (6).

With increasing transmembrane pressure, concentration polarization builds up, and the local concentration close to the membrane increases strongly (evident by the darker colours in figure 5a). When the flux is high enough, a cake layer builds up which reduces the transmembrane flow locally, until almost the complete membrane is covered by a cake layer. At that moment the steady-state transmembrane flux is only determined by the mobility of the particles in the concentration polarization layer, and not on the transmembrane pressure anymore (a higher transmembrane pressure induces a faster supply of particles, build-up of a thicker cake layer, until the flux has reached the same value as before).

Microsieves feature extremely high permeabilities, and therefore already at very low pressures, the critical flux is achieved, after which cake layer formation starts. Thus, controlling the transmembrane pressure sufficiently accurately is not trivial: the pressure needed over the feed channel is of the same order of magnitude. The requirements for the process surrounding the microsieve are therefore much stricter than for conventional membranes.

This already indicates that also fluid behaviour and particle migration need to be taken into account, which will be done in the next section.

Effects related to Fluid Dynamics and Particle Migration

Fluid dynamics and particle migration are normally not considered in great detail during the design of a membrane separation process. We will start this section with an example from the dairy science, showing the importance of these aspects, followed by visualisation of the effects that play a role near the membrane during fractionation. In the last section the effect of different back-transport mechanisms is discussed.

Fractionation of Milk Fat: Effect of Process Conditions

Milk fat globules, which have a broad particle size distribution from 1 to 10 µm, were fractionated with a tubular, ceramic MF membrane with 5.0 µm average pore size. Filtration temperature, cross-flow velocity, wall shear rate and concentration of the suspension were set to and maintained at fixed values (for more information see (7)). The pressure over the membrane was varied, to control the permeate flux. The relation between the permeate flux and the trans-membrane pressure was linear, indicating that no particle accumulation (pore blockage or cake layer formation) was taking place at the membrane.

In Figure 6, the fat content of the permeate relative to the fat content of the milk, and the median of the particle size distribution, are shown as function of the permeate flux for two cross-flow velocities. The particle size distributions in the feed and permeate are different; fractionation takes place. For the highest cross-flow velocities, the particle size and fat content are relatively constant as function of the flux, but much lower than in the feed.

144

For the lower cross-flow velocity, the particle size and fat content clearly increase with higher flux, while the particle size and fat content almost reach the value in the feed solution at the highest flux measured (7).

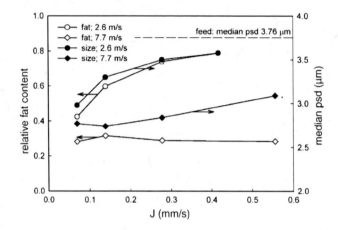

Figure 6. The relative fat content and the median of the particle size distribution in permeate as a function of permeate flux at different cross-flow velocities during MF of milk fat globules. Reproduced with permission from reference 7. Copyright 2006.

As described before, the effects cannot not be related to accumulation of components at the membrane surface, as would be expected for most membrane separation processes. Therefore, the flow conditions have to induce the separation of the suspension. This implies that the composition of the permeate can be controlled through the combination of the cross-flow velocity and the permeate flux (7). The interaction of these effects was further investigated with CSLM analysis as described in the next section.

Fractionation of Latex Suspension: Effect of Particle Size

When investigating the steady-state flux during filtration of bi-dispersed latex suspensions (the latex particles could not pass the membrane), the steady-state flux in the constant-flux regime (i.e, the regime in which the membrane is largely covered with a cake layer) was always dominated by the small particles, irrespective of the ratio in which the particles were mixed (8); see Figure 7.

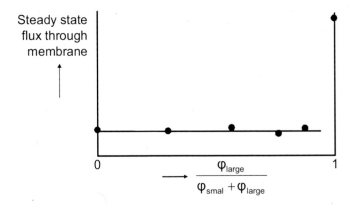

Figure 7. Schematic representation of steady state flux measured for various ratios of small (1.6 µm) and large particles (9.8 µm).

This was a surprising finding, since the mobility of small particles is strongly influenced by the presence of only a small number of larger particles. One would therefore expect that the presence of larger particles would act as mobility enhancers for the smaller ones. Obviously, this does not take place at all. This was investigated further with fluorescent polystyrene particles tracked with a confocal scanning laser microscope in order to study their deposition behavior on the membrane.

A polyethersulfon UF membrane with a molecular cut-off of 100 kDa was used, particle size of the particles was 1.6, 4.0 and 9.8 µm. A module with rectangular channel having fixed dimensions and fixed cross-flow velocity, shear rate, and trans-membrane pressure was used. The total concentration of particles was constant, while the ratio between small and large particles was varied.

In Figure 8, deposition of a bidisperse suspension can be seen during MF at three different times; the feed solution consists of 97.5% (v/v) large particles, while 2.5% is small (7).

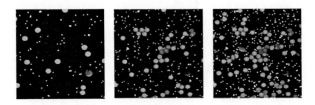

Figure 8. Particle deposition of small (1.6 µm) and large particles (9.8 µm) onto the membrane during MF monitored with CSLM. Time proceeds from left to right. Reproduced with permission from reference 7. Copyright 2006.

146

During filtration, initially deposition of both particles takes place, however, many more small particles deposit relative to the low amount present in the feed and their number increases in time. Apparently, segregation occurs in the liquid, with small particles closer to the membrane wall. This effect is illustrated in Figure 9, where the relative surface area of small particles is plotted as a function of time, for three different ratios of small and large particles. In all cases the surface load of the membrane is completely dominated by the small particles; obviously for lower concentrations of small particles this takes longer (7). We infer from this that segregation of particles occurs already in the fluid above the membrane, which then leads to preferential deposition of smaller particles, even when the overwhelming majority of all particles is large.

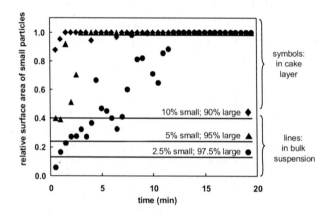

Figure 9. Relative surface load of small particles as a function of time for a bidisperse suspension with 1.6 and 9.8 μm particles.

Effect on Backtransport Mechanisms

From Figures 7-9, one can conclude that the smaller and larger particles segregate in the feed channel. The basis of this behavior may be the nature of the back-transport mechanism. In MF fractionation, three important back-transport mechanisms can be distinguished: Brownian diffusion, shear-induced diffusion, and inertial lift (9). Their relative importance depends on the particle size, the concentration and the viscosity of the suspension. In the cases investigated here, it is most likely that shear-induced diffusion is the major back-transport mechanism, since it is known to be dominant for particle sizes between 1 and 10μm. The shear-induced diffusion coefficient is proportional to γa^2 (10,11). Where is the shear rate γ (s^{-1}) and a is the particle radius (m). Obviously, the large particles are much more influenced by shear induced diffusion given the quadratic relation.

As the name implies, shear-induced diffusion is caused by the shear in the feed channel. Particles pass other particles in slower-moving fluid streamlines and may interact (either by direct collision or by longer-range interaction through disturbance of the surrounding flow field). When these interactions involve three or more particles at some stage, irreversible displacement of the particles from their streamlines results (*12*). This creates a net migration away from the membrane towards the bulk (*13*).

This migration is however not an ideal process that is only dependent on the concentration gradient. It was found to be also dependent on the shear rate gradient and the gradient in viscosity, while a bidisperse suspension was found to show local segregation: larger particles tend to accumulate in the regions with the lowest shear rate, while the smaller ones accumulate in the regions with the higher shear rates (*14*). This effect of shear-induced segregation explains the behaviour seen in our experiments.

It is clear that many aspects play a role in fractionation. It is tempting to focus on the design of a microsieve itself (pore size and shape, etc), but our results show that the fluid dynamics around the membrane have an influence that is as least as large as the design of the membrane itself, not only for optimization of the flux but for the retention as well. In some cases, the fluid dynamics might even be of more importance than the membrane itself, as was shown. Additionaly, in any system, adsorption of particles should be minimized. Therefore, control of the surface properties is an important aspect as well to take into account; this is discussed in chapter XX by Michel Rosso, *et al.*

Conclusion

In the design of a microfiltration process, one needs to simultaneously take into account size scales ranging from colloidal (particles and pore) size to module size.

First, on the scale of pores, the 3D pore geometry strongly influences the force needed to remove a particle from a pore; at larger coverage of the membrane surface by particles, the force needed to remove the particle has to be increased several fold due to shielding effects. At larger scale, the dynamics of suspension particles in the feed channel was found to be complex. Particle deposition on the membrane surface was found to be almost exclusively by smaller particles. This effect was found to result from the dynamics of a polydisperse suspension under shear, i.e., from shear induced segregation, leading accumulation of larger particles in low-shear regions, and expulsion of smaller particles in regions with higher shear (in our case, near the membrane).

Thus, separation efficiency can be increased by changing the dimensions and geometries of the microsieve design, but understanding and using the dynamic behaviour of the feed suspension (especially the particle migration and segregation phenomenon) one can improve the fractionation even without considering the membrane itself. It seems that taking into account the influence of process conditions on the particle behaviour in the whole system, is crucial for significantly increasing fractionation yield of suspensions.

148

References

1. Brans, G.; Schroën, C.G.P.H.; van der Sman, R.G.M.; Boom, R.M. Membrane fractionation of milk: state of the art and challenges. *Journal of Membrane Science* **2004**, 234(1-2), 263-272.
2. van Rijn, C.J.M.; Elwenspoek, M.C. Micro filtration membrane sieve with silicon micro machining for industrial and biomedical applications. *Proceedings of Micro Electro Mechanical Systems (MEMS)* Amsterdam, the Netherlands **1995**, 83.
3. Vogelaar, L.; Lammertink, R.G.H.; Barsema, J.N.; Nijdam, W.; Bolhuis-Versteeg, L.A.M.; van Rijn, C.J.M.; Wessling, M. Phase separation micromolding: a new generic approach for microstructuring various materials. *Small 1* **2005**, 1(6), 645-655.
4. Brans, G.; van der Sman, R.G.M.; Schroën, C.G.P.H.; van der Padt, A.; Boom, R.M. Optimization of the membrane and pore design for micro-machined membranes. *Journal of Membrane Science* **2006**, 278(1-2), 239-250.
5. Brans, G.; Kromkamp, J.; Pek, N.; Gielen, J.; Heck, J.; van Rijn, C.J.M.; van der Sman, R.G.M; Schroën, C.G.P.H.; Boom, R.M. Evaluation of microsieve membrane design. *Journal of Membrane Science* **2006**, 278(1-2), 344-348.
6. Kromkamp, J.; Bastiaanse, A.; Swarts, J.; Brans, G.; van der Sman, R.G.M.; Boom, R.M. A suspension flow model for hydrodynamics and concentration polarisation in crossflow microfiltration. *Journal of Membrane Science* **2005**, 253(1-2), 67-79.
7. Kromkamp, J.; Faber, F.; Schroën, K.; Boom, R. Effects of particle size segregation on crossflow microfiltration performance: Control mechanism for concentration polarisation and particle fractionation. *Journal of Membrane Science* **2006**, 268(2), 189-197.
8. Kromkamp, J.; van Domselaar, M.; Schroën, K.; van der Sman, R.; Boom, R. Shear-induced difftision model for microfiltration of polydisperse suspensions. *Desalination* **2002**, 146, 63-68
9. Samuelsson, G.; Huisman, I.H.; Trägårdh, G.; Paulsson, M. Predicting limiting flux of skim milk in crossflow microfiltration. *Journal of Membrane Science* **1997**, 129(2), 277-281.
10. Karnis, A.; Goldsmit, H.L.; Mason, S.G. Kinetics of Flowing Dispersions. I. Concentrated Suspensions of Rigid Particles. *Journal of Colloid and Interface Science* **1966**, 22(6), 531-553.
11. Lee, Y.; Clark, M.M. Modeling of flux decline during crossflow ultrafiltration of colloidal suspensions. *Journal of Membrane Science* **1998**, 149(2), 181-202.
12. Eckstein, E.C.; Bailey, D.G.; Shapiro, A.H. Self-Diffusion of Particles in Shear-Flow of a Suspension. *Journal of Fluid Mechanics* **1977**, 79 (Jan20), 191-208.

13. Breedveld, V.; van den Ende, D.; Jongschaap, R.; Mellema, J. Shear-induced diffusion and rheology of noncolloidal suspensions: Time scales and particle displacements. *Journal of Chemical Physics* **2001,** 114(13), 5923-5936.

14. Husband, D.M., Mondy, L.A., Ganani, E., Graham, A.L. Direct Measurements of Shear-Induced Particle Migration in Suspensions of Bimodal Spheres. *Rheologica Acta* **1994,** 33(3), 185-192.

Chapter 10

Biorepellent Organic Coatings for Improved Microsieve Filtration

Formation of Covalent Organic Monolayers on Silicon Nitride and Silicon Carbide Surfaces

Michel Rosso,[1,2] Karin Schroën,[2] Han Zuilhof[1]*

[1]Laboratory of Organic Chemistry, Wageningen University, Dreijenplein 8, 6703 HB Wageningen, The Netherlands; [2]Laboratory of Food and Bioprocess Engineering, Wageningen University, Bomenweg 2, 6703 HD Wageningen, The Netherlands; E-mail: Han.Zuilhof@wur.nl

Microsieves are a new type of Si-based membranes, which are coated with a Si_xN_4 top layer. Although Si_xN_4 is known to be relatively inert, surface contamination (fouling through e.g. protein adsorption) is critical for application in microfiltration. As a result, surface modification is needed to prevent or minimize these interactions. Functional coatings can be formed on the Si_xN_4 surfaces via several grafting methods which are presented here. Some stable modifications allow the covalent grafting of biorepelling oligomers and polymers whose effect on protein adsorption is shown.

Microsieves are a new type of micro-perforated filtration membranes, produced using photolithography to create membrane pores with a very well-defined size and shape (1-3). Crystalline silicon is used to form the well-defined 3-D support structure of microsieves by controlled anisotropic etching, while silicon-enriched silicon nitride (Si_xN_4; x typically 3.5 - 5) - deposited by chemical vapor deposition (CVD) - is used as outer coating for its high mechanical and chemical stability (4,5).

Studies have been done to assess and optimize the performance of microsieves in filtration, especially to purify fluids in food processes, like the removal of yeast from beer or the cold sterilization of milk (2). The very thin effective membrane layer of microsieves (< 1 μm) results in a high permeability, and allows very low transmembrane pressures (< 100 mbar (6)) compared to ceramics membranes (with pressures of typically 0.5 to 5 bar (7,8)). Moreover, the possibility to accurately design the pore size and shape, the porosity and the thickness of membranes gives a new freedom to optimize the filtration properties (9-12). However, microsieves suffer, like other microfiltration membranes, from surface contamination, which causes a dramatical decrease in the permeate flux during the filtration (13-15). Such membrane fouling is one of the main limitations for industrial microfiltration (16,17).

For biological solutions in the food, beverage or biotechnology industries, surface contamination is mainly due to proteins or protein aggregates. Protein fouling in microfiltration has been described as a combination of pore blockage, formation of a cake layer on top of the membrane and/or adsorption inside the pores (See Figure 1) (17-19). Mostly, protein adsorption is the first stage of irreversible fouling by other components, therefore tackling protein adsorption may be a solution to a bigger problem.

Figure 1. Sources of membrane fouling during microfiltration.

In this chapter, we present ways to control the surface properties of membranes to prevent the initial adsorption of proteins or other biomaterials. Protein adsorption is strongly reduced on hydrophilic substrates (20,21), because the macromolecules then compete with water to interact with the solid surfaces. In contrast, the use of hydrophilic surfaces may reduce fouling, but cannot totally prevent protein adsorption. To further improve this, strategies involving the grafting of organic monolayers or polymer brushes need to be applied, and we present here such approaches to prevent adsorption onto surfaces.

Besides biorepellance, hydrophilic coatings also provide a good wetting of membrane pores (22), which is especially critical at the low transmembrane pressure and the high product throughput used during microsieve filtration (23).

Surface Properties of Clean Microsieves.

The Si_xN_4 outer coating of microsieves reinforces the spatial structure of the membranes and protects surfaces from corrosion by filtrated solutions, cleaning solutions, and exposure to air during storage. Silicon carbide (SiC) is also a very robust material (24,25) with a high potential in biocompatible devices (26-28).

The possibility to form homogeneous SiC coatings by CVD (29) offers a possible alternative to Si_xN_4 for the coating of microsieves. This chapter will therefore consider the surface modification of both materials.

Properties of Si_xN_4 and SiC Surfaces

Besides silicon, the clean surfaces of pure Si_xN_4 and SiC should contain only nitrogen and carbon, respectively. However, the composition of these surfaces changes upon oxidation and contamination from air, both occurring when the material is stored under ambient conditions. Firstly, the surface of both materials is usually covered with an oxygen-rich top layer after storage in air (24,30). As a result, the oxidized surfaces of Si_xN_4 (30,31) and SiC (32,33) behave very similar to the surface of pure silica, presenting mainly Si-OH groups and a point of zero charge at about pH 3. Consequentely, at physiological pH, these surfaces are negatively charged because of Si-O⁻ surface groups (34,35), and the remaining surface Si-OH groups can also form strong electrostatic dipole interactions and hydrogen bonds with polar compounds, including water. Apart from this oxide layer, the solid surfaces are usually contaminated by organic compounds from the ambient air (36).

When exposed to biological solutions, such oxidized surfaces are quickly covered with a layer of adsorbed proteins or polysaccharides, which in turn favor adhesion of bigger aggregates or microorganisms and biofilm growth (37-39).

Cleaning of Si_xN_4 and SiC Surfaces

A sacrificial oxidation step is needed to obtain surfaces free from organic contamination. The oxidation of Si_xN_4 and SiC substrates can be carried out by chemical treatment with acidic, basic or oxidizing solutions, or their mixtures. The resulting clean oxide layer can be subsequently removed using aqueous solutions of pure or buffered HF. A typical oxidative cleaning procedure involves oxidation in "piranha" solution (H_2SO_4:30% H_2O_2, 3:1), followed by etching with a diluted HF solution (2-5%) (4). Etching with HF leaves the surface of Si_xN_4 deprived of carbon and oxygen (40,41), whereas SiC surfaces still remain oxygen-terminated (42,43). The resulting monolayer of surface hydroxyl groups on SiC cannot be removed unless high-temperature annealing in pure hydrogen or silicon vapor is applied (44,45).

Alternative oxidation methods involve thermal or plasma treatments (40,43,46) in air or oxygen, which combine efficient oxidation with the convenience of dry treatments, and also allow a non-destructive treatment of porous microstructures such as microsieves. Apart from its common application in cleaning and oxidation, plasma treatment is also used to obtain highly hydrophilic organic membranes (21,47) and has as such been applied to Si_xN_4 microsieves (23). This dry process can be easily scaled up from the dimensions of experimental microsieves (5 x 5 mm) to the wafer-sized membranes required for industrial processes. However, the effects of the plasma treatment are only temporary, as the high surface energy of the obtained oxide surface promotes the adsorption of new contaminants from air or water. As a result, the treatment must then be repeated to recover a highly hydrophilic, wettable surface, which again only remains this way for a short period.

Surface Coating of Microsieves

Attachment of Organic Monolayers

The formation of covalent organic monolayers has been widely used to attach specific functional groups onto inorganic substrates. Additionaly, organic monolayers can serve as an anchoring base for further grafting of polymer brushes onto surfaces. The rich chemistry of organic polymers can then be explored to produce microsieves with tailored surface properties.

One of the most common methods to form organic monolayers involves the adsorption of alkylsilanes onto oxidized inorganic surfaces (48,49). These monolayers can be formed by the reaction of chlorosilane or alkoxysilane precursors with any oxidic inorganic material that displays stable hydroxyl groups at their surface (See Figure 2a). This is the case for Si_xN_4 and SiC with a top layer of native silicon dioxide (or, in the case of SiC, also on bare etched surfaces (45)). Oxidation of these materials by controlled chemical or physical treatments (*vide infra*) will then reproducibly provide proper surfaces for the formation of organo-silane monolayers (50-52). However, specifically the hydrolytic stability of such layers is not as high as needed in a practical application where cleaning steps can be very severe. As a result, the potential use of silane-based monolayers is rather limited in membrane applications.

Another method to form stable alkyl monolayers on silicon-containing surfaces involves the reaction of terminal alkenes or alkynes with HF-treated substrates. This method is a variation on the widely used hydrosilylation reaction between 1-alkenes and hydrogen-terminated silicon surfaces (53-57). That reaction occurs through the addition of Si-H on the terminal carbon-carbon double bond, following a thermal initiation (58-60), or a photochemical initiation with UV (61-65) or visible light (66-68). These reactions allow the use of a wider range of reactive moieties that are not compatible with silanes. They also do not need an oxide layer, since they do not require surface -OH groups, but even when hydroxyl-terminated, surfaces can be coated in similar conditions with alkene molecules (43).

Alkene-based monolayers were formed on flat Si_xN_4 (40,41) and 6H-SiC and polycrystalline 3C-SiC (43) using thermal conditions close to those used for the surface modification of silicon. Good quality monolayers were obtained with several simple alkenes (e.g. water contact angles up to 107° for hexadecene-derived monolayers on both SiC and Si_xN_4). One advantage of alkene-based monolayers is their high stability. This is mainly due to the absence of a silicon oxide layer, and to the presence of stable Si-C and Si-N bonds, in the case of Si_xN_4 (40), and extremely stable C-O-C bonds that even survive boiling at pH = 0 for hours, in the case of SiC surfaces (43) (See Figure 2b and c).

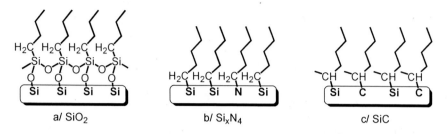

Figure 2: Monolayer formation on Si-based materials using a) organosilanes on oxide surfaces, and alkene-based coatings on b) SixN4 and c) SiC.

Other less general methods can be used to graft small organic molecules or polymer chains onto inorganic surfaces: in particular, surface -NH_2 groups on Si_xN_4 can be reacted with alkyl bromides (69-71) or aldehydes (72). Only a few examples of direct "grafting-on" polymerization to inorganic surfaces exist in the literature. Some reports mention the polymerization of methyl acrylate on Shirazu porous glass (SPG) membranes induced by plasma (73,74), the grafting of vinyl-terminated polystyrene onto silicon oxide surfaces (75) or the electro-grafting of poly-*N*-succinimidylacrylate (PNSA) on SiN cantilevers (76).

Biorepellent Oligomer and Polymer Coatings

Biorepellent coatings are typically hydrophilic, as this strongly minimizes the adsorption of proteins (77-80). In this category two approaches are present, each with their own mode of action. For short oligomers, the high internal hydrophilicity of the grafted chains traps a big amount of water at the liquid-solid interface, and the resulting layer forms a barrier for the adsorption of e.g. colloids from the liquid phase (81). In the case of long polymers, the biorepellance is mainly caused by the osmotic effect of hydrated chains. The adsorption is unfavorable, as the concomitant compression of the grafted chains would locally increase the polymer concentration near the surface (82,83).

Polyethylene glycols (PEG), also called polyethylene oxides (PEO), constitute one of the most widely used polymers to make surfaces biorepellent (84,85). Table 1 presents examples of hydrophilic polymers that have been used to improve the surface properties of membrane. Most of these coatings have been done on organic substrates, but it is possible to extend these modifications to inorganic membranes with a proper grafting method (*vide infra*).

Table 1. Examples of biorepellent hydrophilic polymers.

Monomer	Polymer structure	Reference
Ethylene glycol (EG)		(84,85)
Acrylic acid (AA)		(21,86-88)
Polyethylene glycol methacrylate (PEG-MA)		(86,88,89)
Diethylene glycol vinyl ether (DEGVE)		(90)
Vinyl acetate (VA)		(91)
Vinylsulfonic acid (VSA)		(92)

The grafting of oligomers and polymers onto surfaces can be done using two different approaches: "grafting-on" methods involve the attachment of pre-formed polymer chains onto surfaces, whereas "grafting-from" methods use the *in-situ* polymerization of monomers from the surface (See Figure 3).

When applied to small oligomers, the "grafting-on" approach gives a close packing of the grafted chains and a good passivation of the substrate (93). In addition, the grafted compounds can be precisely characterized before attachment, ensuring a good control of the coating. In particular, the use of organosilanes functionalized with oligoethylene glycols has been reported extensively: chains with 3 to 9 ethylene glycol units are usually sufficient to significantly reduce the adsorption of proteins (94-96) and microorganisms (97) on silica. Organosilane compounds were also used to form monolayers of oligoethylene glycol onto oxidized silicon nitride substrates (98), but no studies on their protein resistance or hydrolytic stability has been reported.

In a recent study from our laboratories monolayers of alkene-based oligoethylene glycol compounds (3 and 6 repeating EG units) were formed on $Si_{3.9}N_4$ substrates (39), which decrease the adsorption of proteins down to the detection limit of reflectometry. Complementary water contact angles and AFM measurements on surfaces coated with compounds containing 6 ethylene glycol units confirmed the absence of any contamination, when tested for BSA and fibrinogen adsorption. Since these monolayers display a very high hydrolytic stability, even in acidic or basic media (40,41), this technique can be applied to microsieves, and studies on this topic are currently ongoing in our laboratories.

Some articles also reported on the attachment of linear PEG polymers attached onto Si_xN_4 cantilevers using polymer chains bearing NHS (N-hydrosuccinimide esters, forming amide bonds with the surface NH_2 groups (99-101), to study molecular recognition with atomic force microscopy. Another example mentions the grafting of PEG chains (MW: 5000) on flat silicon nitride substrates by direct condensation at 100 °C of –OH groups with the silanol groups at the surface to form strong Si-O-C bonds (102). Besides PEG, other linear compounds, such as some derivatives of zwitterionic phospholipids (103,104) ($-(CH_2)_{15}-(PO_4)^- -(CH_2)_2-NR_3^+$ (R = CH_3 or H)) have been attached to oxide surfaces, reducing lysozyme or fibrinogen adsorption by about 90% compared to bare substrates.

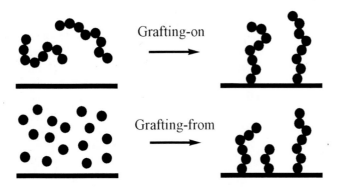

Figure 3: "Grafting-on" and "Grafting-from" strategies for the formation of polymer brushes on solid surfaces.

If a silane group is present as part of the polymer, it can be grafted directly to the surfaces. This modification has been used for instance to attach long PEG chains (MW 750-5000) on silica (105,106) or glass (107) surfaces and on alumina membranes (108). PEG methyl ether acrylate could also be included in a copolymer (109) with 3-(trimethoxysilyl)propyl methacrylate (poly(TMSMA-r-PEGMA), MW of PEG: 475) and attached onto oxidized silicon substrates. These coatings could reduce dramatically the adsorption of BSA an fibrinogen, but could also reduce the adhesion of fibroblast cells onto surfaces. Still using silane chemistry, the pre-formed poly(TMSMA-r-PEGMA) copolymer has also been applied to Si_xN_4 microsieves to introduce PEG chains onto the membrane surfaces (14). Such modified membranes displayed a more efficient filtration of BSA or skimmed milk than untreated microsieves, when using backpulsing. However, these coatings had a relatively low stability at pH 12, which can be related to the instability of silanes in alkaline solutions. Silane pre-coatings also allowed the grafting of amines (110,111) or aldehydes (112), which were then grafted with PEG or dextran, respectively.

Since the "grafting-from" of polymer chains on solid substrates has been developed mainly for organic substrates (113-115) and organic membranes (88,116-119), polymer grafting on Si_xN_4 or SiC surfaces requires the prior formation of organic monolayers. Indeed, using such a method, the polymerization of oligo(ethylene glycol) methyl methacrylate (OEGMA) was carried out by atom transfer radical polymerization (ATRP) (120) and ozone-induced graft polymerization (121), to give protein-repellent glass surfaces. Besides ATRP and ozone initiation, UV irradiation (87,122,123), plasma-induced graft polymerization (21,88,116), or radical initiators such as azobisazobutyronitrile (AIBN) (124) and the redox couple $K_2S_2O_8$ - $Na_2S_2O_5$ (86,89,92) can initiate graft polymerization. Although not all these methods have already been used to graft polymers onto organic monolayers, they represent an important route to exploit the chemistry of polymers on inorganic substrates.

Direct graft-polymerization was also applied in plasma-induced graft polymerization of poly(ethylene glycol) methyl ether methacrylate (PEGMA, MW of PEG: 1000) onto oxidized silicon surfaces; the resulting polymer layers reduced significantly the adhesion of BSA and platelets (125). This direct grafting of polymers onto inorganic surfaces represents a simple alternative to the use of preformed monolayers, but the quality of the grafting is then difficult to control and frequently a low packing density is obtained for entropic reasons. As a result the nature of the polymer-inorganic substrate interface and the possibilities to optimize the layer properties must be investigated in each case.

A possible adverse effect of polymer coatings is a potential increase of membrane resistance, due to the growth of thick layers within the pores. In addition, poorly defined polymerizations may yield polymer layers with irreproducible thicknesses and other properties (stability, protein resistance). However, when properly controlled, the attachment of an organic polymer layer of well-defined thickness can tune small pore sizes, especially in ultrafiltration membranes, and thus in fact turn a potential problem into a potential benefit.

Conclusion

Modification of the surface properties of silicon nitride and silicon carbide substrates can be achieved via various techniques. These can rigorously change the surface properties and biorepellent behavior. The most extensively studied monolayers involve organosilane-based monolayers, either as a fully functional layer or as a reactive intermediate layer from which e.g. grafting-on or grafting-from polymerization can be started. In view of its superior stability, use of milder chemicals, ease of practical use and wider range of allowed functionalities, the more recently developed alkene-based chemistry represents a important improvement as a grafting technique for functional coatings (oligomers, polymers) on silicon nitride and silicon carbide.

Using the wide range of grafting techniques and available biorepellent compounds, surface chemistry can provide efficient ways to improve filtration with microsieves and thus expand their range of applications.

ACKNOWLEDGMENT. The authors thank Graduate School VLAG and MicroNed (Project no. 6163510395) for financial support.

References

(1) van Rijn, C. J. M.; Nijdam, W.; Kulper, S.; Veldhuis, G. J.; van Wolferen, H.; Elwenspoek, M. *J. Micromech. Microeng.* **1999**, *9*, 170-172.

(2) Van Rijn, C. J. M., *Nano and Micro Engineered Membrane Technology.* Aquamarijn Research BV, The Netherlands: 2002.

(3) Kuiper, S.; van Wolferen, H.; van Rijn, G.; Nijdam, W.; Krijnen, G.; Elwenspoek, M. *J. Micromech. Microeng.* **2001**, *11*, 33-37.

(4) Bermudez, V. M.; Perkins, F. K. *Appl. Surf. Sci.* **2004**, *235*, 406-419.

(5) Rathi, V. K.; Gupta, M.; Agnihotri, O. P. *Microelectron. J.* **1995**, *26*, 563.

(6) Brans, G.; Kromkamp, J.; Pek, N.; Gielen, J.; Heck, J.; van Rijn, C. J. M.; Van der Sman, R. G. M.; Schroen, C. G. P. H.; Boom, R. M. *J. Membr. Sci.* **2006**, *278*, 344-348.

(7) Brans, G.; Schroen, C. G. P. H.; van der Sman, R. G. M.; Boom, R. M. *J. Membr. Sci.* **2004**, *243*, 263-272.

(8) Daufin, G.; Escudier, J. P.; Carrere, H.; Berot, S.; Fillaudeau, L.; Decloux, M. *Food Bioprod. Process.* **2001**, *79*, 89-102.

(9) Brans, G.; Van der Sman, R. G. M.; Schroen, C. G. P. H.; van der Padt, A.; Boom, R. M. *J. Membr. Sci.* **2006**, *278*, 239-250.

(10) Kuiper, S.; Brink, R.; Nijdam, W.; Krijnen, G. J. M.; Elwenspoek, M. C. *J. Membr. Sci.* **2002**, *196*, 149-157.

(11) Chandler, M.; Zydney, A. *J. Membr. Sci.* **2006**, *285*, 334-342.

(12) Kuiper, S.; van Rijn, C.; Nijdam, W.; Raspe, O.; van Wolferen, H.; Krijnen, G.; Elwenspoek, M. *J. Membr. Sci.* **2002**, *196*, 159-170.

(13) Mulder, M., *Basic Principles of Membrane Technology.* 2 ed.; Kluver Academic Publishers: Dordrecht, 1996; p 418.

(14) Girones, M.; Bolhuis-Versteeg, L.; Lammertink, R.; Wessling, M. *J. Colloid Interf. Sci.* **2006**, *299*, 831-840.

(15) Girones, M.; Lammertink, R. G. H.; Wessling, M. *J. Membr. Sci.* **2006**, *273*, 68-76.

(16) Belfort, G.; Davis, R. H.; Zydney, A. L. *J. Membr. Sci.* **1994**, *96*, 1-58.

(17) Palacio, L.; Ho, C. C.; Pradanos, P.; Hernandez, A.; Zydney, A. L. *J. Membr. Sci.* **2003**, *222*, 41-51.

(18) Bowen, W. R.; Calvo, J. I.; Hernandez, A. *J. Membr. Sci.* **1995**, *101*, 153.

(19) Ho, C. C.; Zydney, A. L. *J. Colloid Interf. Sci.* **2000**, *232*, 389-399.

(20) Koehler, J. A.; Ulbricht, M.; Belfort, G. *Langmuir* **1997**, *13*, 4162-4171.

(21) Ulbricht, M.; Belfort, G. *J. Membr. Sci.* **1996**, *111*, 193-215.

(22) Franken, A. C. M.; Nolten, J. A. M.; Mulder, M. H. V.; Bargeman, D.; Smolders, C. A. *J. Membr. Sci.* **1987**, *33*, 315-328.

(23) Girones, M.; Borneman, Z.; Lammertink, R. G. H.; Wessling, M. *J. Membr. Sci.* **2005**, *259*, 55-64.

(24) Choyke, W. J.; Matsunami, H.; Pensl, G., *Silicon Carbide, Recent Major Advances.* Springer: Berlin, 2003.

(25) Saddow, S. E.; Agarwal, A., *Advances in Silicon Carbide: Processing and Applications.* Artech House Inc.: Boston, 2004.

(26) Cogan, S. F.; Edell, D. J.; Guzelian, A. A.; Liu, Y. P.; Edell, R. *J. Biomed. Mater. Res. A* **2003**, *67A*, 856-867.

160

(27) Rosenbloom, A. J.; Sipe, D. M.; Shishkin, Y.; Ke, Y.; Devaty, R. P.; Choyke, W. J. *Biomed. Microdev.* **2004,** *6,* 261-267.

(28) Sella, C.; Martin, J. C.; Lecoeur, J.; Lechanu, A.; Harmand, M. F.; Naji, A.; Davidas, J. P. *Mater. Sci. Eng. A* **1991,** *139,* 49-57.

(29) Roper, C. S.; Radmilovic, V.; Howe, R. T.; Maboudian, R. *J. Electrochem. Soc.* **2006,** *153,* C562-C566.

(30) Bousse, L. J.; Mostarshed, S.; Hafeman, D. *Sensor Actuator B Chem.* **1992,** *10,* 67-71.

(31) Harame, D. L.; Bousse, L. J.; Shott, J. D.; Meindl, J. D. *IEEE T. Electron Dev.* **1987,** *34,* 1700-1707.

(32) Popping, B.; Deratani, A.; Sebille, B.; Desbois, N.; Lamarche, J. M.; Foissy, A. *Colloid Surface* **1992,** *64,* 125-133.

(33) Whitman, P. K.; Feke, D. L. *J. Am. Ceram. Soc.* **1988,** *71,* 1086-1093.

(34) Bolt, G. H. *J. Phys. Chem.* **1957,** *61,* 1166-1169.

(35) Hiemstra, T.; Dewit, J. C. M.; Vanriemsdijk, W. H. *J. Colloid Interf. Sci.* **1989,** *133,* 105-117.

(36) Patton, S. T.; Eapen, K. C.; Zabinski, J. S. *Tribol. Int.* **2001,** *34,* 481-491.

(37) Marshall, A. D.; Munro, P. A.; Tragardh, G. *Desalination* **1993,** *91,* 65.

(38) Vanloosdrecht, M. C. M.; Lyklema, J.; Norde, W.; Zehnder, A. J. B. *Microbiol. Rev.* **1990,** *54,* 75-87.

(39) Rosso, M.; de Jong, E.; Giesbers, M.; Fokkink, R. G.; Norde, W.; Schroen, K.; Zuilhof, H., submitted.

(40) Arafat, A.; Giesbers, M.; Rosso, M.; Sudhölter, E. J. R.; Schroen, K.; White, R. G.; Yang, L.; Linford, M. R.; Zuilhof, H. *Langmuir* **2007,** *23,* 6233.

(41) Arafat, A.; Schroën, K.; de Smet, L. C. P. M.; Sudhölter, E. J. R.; Zuilhof, H. *J. Am. Chem. Soc.* **2004,** *126,* 8600-8601.

(42) King, S. W.; Nemanich, R. J.; Davis, R. F. *J. Electrochem. Soc.* **1999,** *146,* 1910-1917.

(43) Rosso, M.; Arafat, A.; Schroen, K.; Giesbers, M.; Roper, C. S.; Maboudian, R.; Zuilhof, H. *Langmuir* **2008,** *24,* 4007-4012.

(44) Bernhardt, J.; Schardt, J.; Starke, U.; Heinz, K. *Appl. Phys. Lett.* **1999,** *74,* 1084-1086.

(45) Starke, U. *Phys. Stat. Sol. B* **1997,** *202,* 475-499.

(46) Chen, L.; Guy, O. J.; Pope, G.; Teng, K. S.; Maffeis, T.; Wilks, S. P.; Mawby, P. A.; Jenkins, T.; Brieva, A.; Hayton, D. J. *Mater. Sci. Forum* **2004,** *457-460,* 1337-1340.

(47) Ulbricht, M.; Belfort, G. *J. Appl. Polymer Sci.* **1995,** *56,* 325-343.

(48) Onclin, S.; Ravoo, B. J.; Reinhoudt, D. N. *Angew. Chem. Int. Ed.* **2005,** *44,* 6282-6304.

(49) Sagiv, J. *J. Am. Chem. Soc.* **1980,** *102,* 92-98.

(50) Petoral, R. M.; Yazdi, G. R.; Spetz, A. L.; Yakimova, R.; Uvdal, K. *Appl. Phys. Lett.* **2007,** *90,* -.

(51) Sampathkumaran, U.; De Guire, M. R.; Heuer, A. H.; Niesen, T.; Bill, J.; Aldinger, F. *Ceram. Trans.* **1999,** *94,* 307-318.

(52) Schoell, S. J.; Hoeb, M.; Sharp, I. D.; Steins, W.; Eickhoff, M.; Stutzmann, M.; Brandt, M. S. *Appl. Phys. Lett.* **2008,** *92,* -.

(53) Boukherroub, R. *Curr. Opin. Solid. State Mater. Sci.* **2005,** *9,* 66-72.

(54) Buriak, J. M. *Chem. Rev.* **2002,** *102,* 1271-1308.

(55) Linford, M. R.; Fenter, P.; Eisenberger, P. M.; Chidsey, C. E. D. *J. Am. Chem. Soc.* **1995**, *117*, 3145-3155.

(56) Shirahata, N.; Hozumi, A.; Yonezawa, T. *Chem. Rec.* **2005**, *5*, 145-159.

(57) Sieval, A. B.; Linke, R.; Zuilhof, H.; Sudhölter, E. J. R. *Adv. Mater.* **2000**, *12*, 1457-1460.

(58) Linford, M. R.; Chidsey, C. E. D. *J. Am. Chem. Soc.* **1993**, *115*, 12631.

(59) Scheres, L.; Arafat, A.; Zuilhof, H. *Langmuir* **2007**, *23*, 8343-8346.

(60) Sieval, A. B.; Demirel, A. L.; Nissink, J. W. M.; Linford, M. R.; van der Maas, J. H.; de Jeu, W. H.; Zuilhof, H.; Sudhölter, E. J. R. *Langmuir* **1998**, *14*, 1759-1768.

(61) Cicero, R. L.; Linford, M. R.; Chidsey, C. E. D. *Langmuir* **2000**, *16*, 5688.

(62) Effenberger, F.; Gotz, G.; Bidlingmaier, B.; Wezstein, M. *Angew. Chem. Int. Ed.* **1998**, *37*, 2462-2464.

(63) Mischki, T. K.; Donkers, R. L.; Eves, B. J.; Lopinski, G. P.; Wayner, D. D. M. *Langmuir* **2006**, *22*, 8359-8365.

(64) Strother, T.; Cai, W.; Zhao, X. S.; Hamers, R. J.; Smith, L. M. *J. Am. Chem. Soc.* **2000**, *122*, 1205-1209.

(65) Strother, T.; Hamers, R. J.; Smith, L. M. *Nucleic Acids Res.* **2000**, *28*, 3535.

(66) de Smet, L. C. P. M.; Pukin, A. V.; Sun, Q. Y.; Eves, B. J.; Lopinski, G. P.; Visser, G. M.; Zuilhof, H.; Sudhölter, E. J. R. *Appl. Surf. Sci.* **2005**, *252*, 24-30.

(67) de Smet, L. C. P. M.; Stork, G. A.; Hurenkamp, G. H. F.; Sun, Q. Y.; Topal, H.; Vronen, P. J. E.; Sieval, A. B.; Wright, A.; Visser, G. M.; Zuilhof, H.; Sudhölter, E. J. R. *J. Am. Chem. Soc.* **2003**, *125*, 13916-13917.

(68) Sun, Q.-Y.; de Smet, L. C. P. M.; van Lagen, B.; Giesbers, M.; Thune, P. C.; van Engelenburg, J.; de Wolf, F. A.; Zuilhof, H.; Sudhölter, E. J. R. *J. Am. Chem. Soc.* **2005**, *127*, 2514-2523.

(69) Cattaruzza, F.; Cricenti, A.; Flamini, A.; Girasole, M.; Longo, G.; Mezzi, A.; Prosperi, T. *J. Mater. Chem.* **2004**, *14*, 1461-1468.

(70) Cricenti, A.; Longo, G.; Luce, M.; Generosi, R.; Perfetti, P.; Vobornik, D.; Margaritondo, G.; Thielen, P.; Sanghera, J. S.; Aggarwal, I. D.; Miller, J. K.; Tolk, N. H.; Piston, D. W.; Cattaruzza, F.; Flamini, A.; Prosperi, T.; Mezzi, A. *Surf. Sci.* **2003**, *544*, 51-57.

(71) Karymov, M. A.; Kruchinin, A. A.; Tarantov, Y. A.; Balova, I. A.; Remisova, L. A.; Vlasov, Y. G. *Sensor Actuator B Chem.* **1995**, *29*, 324-327.

(72) Yin, L. T.; Chou, J. C.; Chung, W. Y.; Sun, T. P.; Hsiung, S. K. *IEEE T. Bio-Med. Eng.* **2001**, *48*, 340-344.

(73) Kai, T.; Suma, Y.; Ono, S.; Yamaguchi, T.; Nakao, S. I. *J. Polym. Sci. Pol. Chem.* **2006**, *44*, 846-856.

(74) Kai, T.; Yamaguchi, T.; Nakao, S. *Ind. Eng. Chem. Res.* **2000**, *39*, 3284.

(75) Maas, J. H.; Stuart, M. A. C.; Sieval, A. B.; Zuilhof, H.; Sudhölter, E. J. R. *Thin Solid Films* **2003**, *426*, 135-139.

(76) Gabriel, S.; Jerome, C.; Jerome, R.; Fustin, C. A.; Pallandre, A.; Plain, J.; Jonas, A. M.; Duwez, A. S. *J. Am. Chem. Soc.* **2007**, *129*, 8410-+.

(77) Kane, R. S.; Deschatelets, P.; Whitesides, G. M. *Langmuir* **2003**, *19*, 2388.

(78) Senaratne, W.; Andruzzi, L.; Ober, C. K. *Biomacromolecules* **2005**, *6*, 2427-2448.

(79) Ramsden, J. J. *Chem. Soc. Rev.* **1995**, *24*, 73-78.

(80) Prime, K. L.; Whitesides, G. M. *Science* **1991**, *252*, 1164-1167.

162

(81) Wang, R. L. C.; Kreuzer, H. J.; Grunze, M. *J. Phys. Chem. B* **1997**, *101*, 9767-9773.

(82) Halperin, A. *Langmuir* **1999**, *15*, 2525-2533.

(83) Jeon, S. I.; Lee, J. H.; Andrade, J. D.; Degennes, P. G. *J. Colloid Interf. Sci.* **1991**, *142*, 149-158.

(84) Chan, Y. H. M.; Schweiss, R.; Werner, C.; Grunze, M. *Langmuir* **2003**, *19*, 7380-7385.

(85) Norde, W.; Gage, D. *Langmuir* **2004**, *20*, 4162-4167.

(86) Freger, V.; Gilron, J.; Belfer, S. *J. Membr. Sci.* **2002**, *209*, 283-292.

(87) Taniguchi, M.; Belfort, G. *J. Membr. Sci.* **2004**, *231*, 147-157.

(88) Ulbricht, M.; Matuschewski, H.; Oechel, A.; Hicke, H. G. *J. Membr. Sci.* **1996**, *115*, 31-47.

(89) Belfer, S.; Purinson, Y.; Fainshtein, R.; Radchenko, Y.; Kedem, O. *J. Membr. Sci.* **1998**, *139*, 175-181.

(90) Bremmell, K. E.; Kingshott, P.; Ademovic, Z.; Winther-Jensen, B.; Griesser, H. J. *Langmuir* **2006**, *22*, 313-318.

(91) Kim, M.; Saito, K.; Furusaki, S.; Sugo, T.; Okamoto, J. *J. Membr. Sci.* **1991**, *56*, 289-302.

(92) Belfer, S.; Purinson, Y.; Kedem, O. *Acta Polymerica* **1998**, *49*, 574-582.

(93) Papra, A.; Gadegaard, N.; Larsen, N. B. *Langmuir* **2001**, *17*, 1457-1460.

(94) Cecchet, F.; De Meersman, B.; Demoustier-Champagne, S.; Nysten, B.; Jonas, A. M. *Langmuir* **2006**, *22*, 1173-1181.

(95) Hoffmann, C.; Tovar, G. E. M. *J. Colloid Interf. Sci.* **2006**, *295*, 427-435.

(96) Lee, S. W.; Laibinis, P. E. *Biomaterials* **1998**, *19*, 1669-1675.

(97) Finlay, J. A.; Krishnan, S.; Callow, M. E.; Callow, J. A.; Dong, R.; Asgill, N.; Wong, K.; Kramer, E. J.; Ober, C. K. *Langmuir* **2008**, *24*, 503-510.

(98) Cerruti, M.; Fissolo, S.; Carraro, C.; Ricciardi, C.; Majumdar, A.; Maboudian, R. *Langmuir* **2008**, ASAP article.

(99) Ebner, A.; Wildling, L.; Kamruzzahan, A. S. M.; Rankl, C.; Wruss, J.; Hahn, C. D.; Holzl, M.; Zhu, R.; Kienberger, F.; Blaas, D.; Hinterdorfer, P.; Gruber, H. J. *Bioconjugate Chem.* **2007**, *18*, 1176-1184.

(100) Riener, C. K.; Stroh, C. M.; Ebner, A.; Klampfl, C.; Gall, A. A.; Romanin, C.; Lyubchenko, Y. L.; Hinterdorfer, P.; Gruber, H. J. *Anal. Chim. Acta* **2003**, *479*, 59-75.

(101) Wang, T.; Xu, J. J.; Qiu, F.; Zhang, H. D.; Yang, Y. L. *Polymer* **2007**, *48*, 6170-6179.

(102) Suo, Z. Y.; Arce, F. T.; Avci, R.; Thieltges, K.; Spangler, B. *Langmuir* **2006**, *22*, 3844-3850.

(103) Wang, Y. L.; Su, T. J.; Green, R.; Tang, Y. Q.; Styrkas, D.; Danks, T. N.; Bolton, R.; Liu, J. R. *Chem. Comm.* **2000**, 587-588.

(104) Feng, W.; Zhu, S. P.; Ishihara, K.; Brash, J. L. *Langmuir* **2005**, *21*, 5980.

(105) Sharma, S.; Johnson, R. W.; Desai, T. A. *Appl. Surf. Sci.* **2003**, *206*, 218.

(106) Sharma, S.; Johnson, R. W.; Desai, T. A. *Langmuir* **2004**, *20*, 348-356.

(107) Yang, Z. H.; Galloway, J. A.; Yu, H. U. *Langmuir* **1999**, *15*, 8405-8411.

(108) Popat, K. C.; Mor, G.; Grimes, C.; Desai, T. A. *J. Membr. Sci.* **2004**, *243*, 97-106.

(109) Jon, S. Y.; Seong, J. H.; Khademhosseini, A.; Tran, T. N. T.; Laibinis, P. E.; Langer, R. *Langmuir* **2003**, *19*, 9989-9993.

(110) Heyes, C. D.; Kobitski, A. Y.; Amirgoulova, E. V.; Nienhaus, G. U. *J. Phys. Chem. B* **2004**, *108*, 13387-13394.

(111) Massia, S. P.; Stark, J.; Letbetter, D. S. *Biomaterials* **2000**, *21*, 2253-2261.

(112) Schlapak, R.; Pammer, P.; Armitage, D.; Zhu, R.; Hinterdorfer, P.; Vaupel, M.; Fruhwirth, T.; Howorka, S. *Langmuir* **2006**, *22*, 277-285.

(113) Uyama, Y.; Kato, K.; Ikada, Y. *Adv. Polym. Sci.* **1998**, *137*, 1-39.

(114) Kato, K.; Uchida, E.; Kang, E. T.; Uyama, Y.; Ikada, Y. *Prog. Polym. Sci.* **2003**, *28*, 209-259.

(115) Hilal, N.; Ogunbiyi, O. O.; Miles, N. J.; Nigmatullin, R. *Separ. Sci. Tech.* **2005**, *40*, 1957-2005.

(116) Chen, H.; Belfort, G. *J. Appl. Polymer Sci.* **1999**, *72*, 1699-1711.

(117) Kilduff, J. E.; Mattaraj, S.; Zhou, M. Y.; Belfort, G. *J. Nanopart. Res.* **2005**, *7*, 525-544.

(118) Yamagishi, H.; Crivello, J. V.; Belfort, G. *J. Membr. Sci.* **1995**, *105*, 237.

(119) Yamagishi, H.; Crivello, J. V.; Belfort, G. *J. Membr. Sci.* **1995**, *105*, 249.

(120) Ma, H. W.; Li, D. J.; Sheng, X.; Zhao, B.; Chilkoti, A. *Langmuir* **2006**, *22*, 3751-3756.

(121) Beyer, M.; Felgenhauer, T.; Bischoff, F. R.; Breitling, F.; Stadler, V. *Biomaterials* **2006**, *27*, 3505-3514.

(122) Ma, H. M.; Bowman, C. N.; Davis, R. H. *J. Membr. Sci.* **2000**, *173*, 191.

(123) Susanto, H.; Balakrishnan, M.; Ulbricht, M. *J. Membr. Sci.* **2007**, *288*, 157-167.

(124) Asatekin, A.; Kang, S.; Elimelech, M.; Mayes, A. M. *J. Membr. Sci.* **2007**, *298*, 136-146.

(125) Zou, X. P.; Kang, E. T.; Neoh, K. G. *Plasmas Polym.* **2002**, *7*, 151-170.

Chapter 11

Dendritic Polymer Networks: A New Class of Nano-Domained Environmentally Benign Antifouling Coatings

Abhijit Sarkar,[1,2] Joseph Rousseau,[1,2] Claire Hartmann-Thompson,[2] Chris Maples,[3] Joe Parker,[3] Paul Joyce,[4] John I. Scheide[5] and Petar R. Dvornic[1,2,*]

[1.] Dendritech, Inc., 3110 Schuette Dr., Midland, MI, 48642;
[2.] Michigan Molecular Institute, 1910 West St. Andrews Rd, Midland, MI 48640;
[3.] Gougeon Brothers, Inc., 100 Patterson Ave., Bay City, MI, 48707;
[4.] Sea Education Association, 171 Woods Hole Rd., Falmouth, MA, 02540;
[5.] Department of Biology, Central Michigan University, Mount Pleassant, MI, 48859
* Phone: +1-989-832-5555 ext. 550; email: dvornic@mmi.org

A new type of antifouling coatings for application in both fresh water and marine environments has been developed utilizing dendritic polymer nanotechnology. The resulting coatings are capable of encapsulating and strongly binding electrophilic biocides in their dendritic nano-cells, where biocides retain their antifouling activity while being prevented from leaching and polluting the aquatic environment. These coatings for the first time successfully combine the best properties of non-stick and biocide-containing coatings and provide an environment-ally benign solution for the protection of man-made objects.

Introduction

Adhesion of aquatic organisms (including microorganisms, plants and animals) to various man-made substrates represents the major component of biological fouling (biofouling) that is a highly undesirable phenomenon occurring on numerous artificial surfaces that are either submerged or in prolonged contact with fresh or sea water. Such surfaces include the hulls of boats and ships, architectural objects such as piers and shore defenses, ship ballast water tanks, components of water purification and production plants, storage containers, pipes and turbines in hydroelectric plants, as well as various objects used in transportation, fishing, underwater drilling, pearl harvesting, defense and numerous other water-based industries. Because biofouling affects all of these, its total cost to governments, industry and individuals are practically impossible to estimate, but they certainly range in the tens of billions of dollars since just the current global market for antifouling paints for the protection of ship hulls is estimated at around $1 billion annually (1).

In operations such as waste water purification and land-based potable water production, biofouling in fresh waters is of predominant concern and it is usually caused by algae (soft fouling) and zebra, *Dreissena polymorpha*, or quagga mussels, *Dreissena bugensis* (hard fouling), while in sea waters (including desalination) the main threats are most often algae and barnacles. Regardless of the nature of the aquatic environment, however, among potential methods considered for solving the biofouling problem, antifouling paints containing dispersed biocides, including heavy metal agents such as organo-copper and especially organo-tin compounds (mostly in the form of tri-*n*-butyl tin, TBT), became very popular in the late 1970s and 1980s because of their extreme effectiveness. Unfortunately, many of these compounds not only eliminate the fouling organisms from the surfaces but their slow release into the water (even in concentrations of only parts per billion) has serious toxic effects on a wide range of other aquatic species, including mollusks, such as oysters and whelks, fish, vegetation and even mammals (including dolphins) (2,3). As a consequence, serious environmental restrictions on paints containing biocides that can leach into the water were introduced in the legislations of many countries. For example, as of January 1, 2003, application of TBT is officially outlawed in the U.S. (2).

For all of these reasons, research interest in recent years has been focused on environmentally benign solutions for biofouling. Among others, targeted goals include non-biocidal foulant-release coatings, fundamental understanding of bioattachment and its basic mechanism (both molecular and cellular), identification of natural repellents, chemical and structural characterization of the surfaces that have non-fouling properties, nano-texturing of submerged surfaces, etc. In the field of foulant-release paints/coatings, the focus has been on low surface energy coatings with very smooth surfaces and external lubricating characteristics (4-7).

Nano-Domained Dendritic Polymer-Based Antifouling Coatings

Among potential candidates for such non-stick antifouling surfaces, silicones and silicon-containing polymers have shown particularly promising characteristics because they possess excellent combinations of some of the most desirable physical properties required for "easy release" (4-7). These properties include: high flexibility (i.e., low glass transition temperatures, Tg, which, for example, for polydimethylsiloxane, PDMS, $-(SiMe_2O)_n-$, the parent polymer of the silicone family, is $-125°C$, among the lowest known to polymer science) (8,14), low surface energy (for PDMS generally between 20 and 28 mN/m; 18-21 dynes/cm^2), relatively high contact angles against water of 99-101° (9,10), and a critical surface tension that coincides with the minimum of a plot of relative attachment versus surface free energy (10). In addition, these polymers are bio-compatible and widely used in cosmetics, the food-processing industry, and for medicinal purposes (10), and even their long-term thermo-oxidative decomposition (14) mainly releases H_2O, SiO_2, and CO_2 in relatively low amounts compared to organic carbon-hydrogen substances, so that PDMS is also generally accepted as an environmentally non-hazardous material. It is because of this exceptional combination of properties that many types of silicon-containing polymers are widely used for different environmentally safe coating applications, including antifouling.

However, although low energy surfaces certainly make it more difficult for aquatic organisms to attach; in order to be truly effective they also need to have repelling characteristics that could influence the foulants to "look elsewhere" for more "friendly" substrates on which to settle. Clearly, such repellents would have to be "unpleasant" to the fouling organisms but either completely natural, or synthetic but biodegradable, non-toxic and not dangerous in any other way. Hence, an ideal solution for the biofouling problem would be a non-stick silicone-based coating that is tough, durable, easy-to-apply, and capable of containing *and retaining without leaching* selected antifouling agents that can effectively repel aquatic nuisance organisms in an environmentally safe way and for a substantial period of time.

Here, we describe a new type of nano-structured coatings that seems to ideally fit this description. As illustrated in Figure 1, these coatings combine two main features: (a) nano-scaled, dendritic polymer cells organized in a precise, 3D honeycomb-like configuration (presented in Figure 1 as circular sections connected in a network) and capable of *chemically binding and physically encapsulating* various guest species, including biofoulant repelling agents (shown as individual dots entrapped within the dendritic cells), with (b) the above described highly desirable methylsilicone low energy surfaces (11,12). Since in these coatings, the encapsulated antifoulants are positioned right at the outside coating surface (i.e., within nanometer thin layers from the water interface) they are able to fully exert their activity across the interface (11,13) and effectively repel aquatic nuisance organisms. However, since they are also firmly bound (i.e., chemically complexed) and physically entrapped (i.e., sterically hindered) inside the highly branched dendritic nano-cells, they are fixed in place and not able to migrate across the interface and into the surrounding water. As a consequence, these unique dendritic polymer-based

168

nano-structured coatings provide effective environmentally benign antifouling protection preventing dangerous pollution while employing proven bio-repellants in combination with smooth, non-stick silicone-based surfaces. To the best of our knowledge, this is the first time that such an effective combination of these most desired properties for antifouling applications has been accomplished.

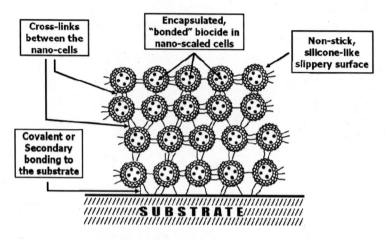

Figure 1: *Schematic representation of the cellular (i.e., honeycomb-like) nano-domained antifouling dendritic polymer coatings. In coatings described in this work, dendritic domains (circular sections) are of either polyamidoamine (PAMAM) or polyurea (PU) compositions and they act as nano-sized containers capable of chemically complexing and physically encapsulating functionally active antifoulants represented by small dots inside the circular sections. Note that the outer coating surface (top) has organo-silicon (OS) composition and its corresponding (e.g., non-stick) physico-chemical characteristics.*

Dendritic Polymers

Dendritic polymers are highly branched macromolecules that are based on characteristic "branch-upon-branch-upon-branch" or "cascade-branched" structural motifs (15,16). These motifs represent the single structural feature that distinguishes dendritic molecules from all other types of macromolecular architectures, including traditional branched polymers, where branching is generally occasional (i.e., primary branches, rarely secondary branches and very rarely tertiary branches) and is sometimes even present as an imperfection rather than the defining and repeating structural feature. In principle, dendritic polymers comprise five major subgroups, including: dendrimers, hyperbranched polymers, dendrons, dendrigrafts or arborescent polymers, and dendronized polymers (16).

Dendrimers (17) (see Figure 2A) are globular macromolecules that consist of two or more tree-like dendrons emanating from either a single central atom or atomic group called the core. They are built of three-dimensional repeat units that contain at least one branch juncture, and in an idealized case of complete and perfect connectivity (i.e., in an ideal dendrimer structure) these repeat units are organized in a series of regular, radially concentric layers (called generations) around the core. Each of these layers contains a mathematically precise number of branches which increases in a geometrically progressive manner from the core to the dendrimer exterior (often also referred to as the dendrimer surface). As a consequence, dendrimers are highly regular, spheroidal molecules, characterized with very low polydispersity and their outermost branches end with end-groups that can be either chemically inert or reactive. The number of dendrimer end-groups depends on the particular dendrimer composition (i.e., the branching functionality at each generational level) and generation, and may range from only a few (i.e., 3 or 4 functional groups of common cores) to several hundreds or even thousands at very high generations (16). Because of this, high generation dendrimers may be viewed as globular, reactive nanoscopic building blocks with some of the highest density of functionality known to chemistry.

Similar to dendrimers, hyperbranched polymers (see Figure 2B), also consist of tree-like macromolecules with extensive intramolecular branching and large numbers of end-groups (denoted B in Figure 2B) (18). However, their structures are less regular and symmetrical than those of dendrimers and they exhibit statistical size and shape distributions that can in some cases be very broad with polymolecularity coefficients ranging to over 15. On the other hand, hyperbranched polymers are also significantly more economical and easier to prepare by various one-step, one-pot procedures than dendrimers. As a consequence, hyperbranched polymers have attracted considerable scientific attention as potential substitutes for dendrimers particularly in applications that can tolerate their polymolecularity, including functional and protective coatings to create antifouling surfaces. In this work, we evaluated both types of polymers for the preparation of nano-domained antifouling coatings, focusing, however, on the utility of hyperbranched polymers while using the compositionally closely related dendrimers as model compounds for the proof of principle studies.

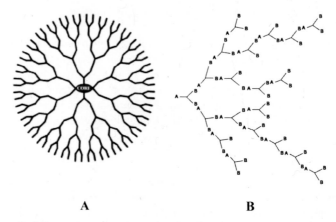

A   B

Figure 2: *Schematic representation of molecular branching topology of a tetradendron dendrimer **A** (i.e., a dendrimer with four dendrons emanating from a single central core) and a hyperbranched polymer molecule **B** resulting from a step-growth polymerization of an AB₂ monomer. Note that in the hyperbranched structure the Bs represent the end-groups which are not lettered in the dendrimer representation.*

Silicon-Containing Dendritic Polymers for Nano-Domained Antifouling Coatings

In order to combine the desired properties of silicon-containing and dendritic polymers summarized in the preceding sections (19), we recently developed several new types of radially layered poly(amidoamine-organosilicon), PAMAMOS, dendrimers (see Reaction Scheme 1) (20-22) and their less perfect but closely related hyperbranched analogues, poly(urea-siloxanes), HB-PUSOX (see Reaction Scheme 2) (23,24). Compositionally, both of these unique unimolecular, inverted micelle-type nano-compounds are comprised of hydrophilic and nucleophilic polyamidoamine, PAMAM, or polyurea, PU, interiors and hydrophobic organosilicon, OS, exteriors (20-24). Architecturally, because of the globular shapes and nanoscopic sizes of their multi-functional macromolecules (which range in diameters, depending on the generation or molecular weight, between about 1 and 10 nm), these unique dendritic polymers represent excellent precursors (i.e., building blocks) for a controlled preparation of a variety of different types (see Figure 3) of covalently crosslinked, nano-domained networks of Figure 1, comprised of 3D arrays of nano-scaled PAMAM or PU cells and OS domains, as shown in Reaction Schemes 1 and 2 (11,25). These networks can be readily processed into mechanically tough and resistant, hard-to-scratch films, sheets and coatings on a variety of substrates, including metals (such as steel or aluminum), glass fiber-reinforced plastics, glass, cement, etc. (11,12,26).

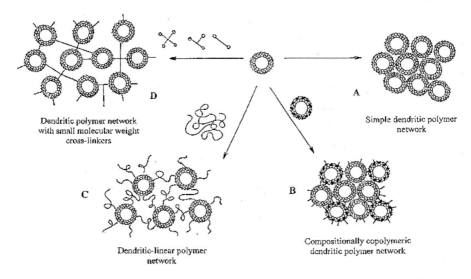

Figure 3: *Different types of honeycomb-like, nano-domained dendritic polymer networks that can be obtained by appropriate combinations of dendritic and non-dendritic co-reagents. Variations in structure enable control of the mechanical properties of the resulting coatings.*

Furthermore, the unique copolymeric composition of PAMAMOS dendrimers and HB-PUSOX hyperbranched polymers makes them exceptionally effective nano-containers for absorption, complexation, encapsulation *and retention* of various guest species. Their complexing power is provided by the nucleophilic character of their PAMAM, $[-(CH_2)_2-C(O)N(H)-(CH_2)_2-N<]$, and urea, $\{-C(O)N(H)-CH_2-C_6H_7(CH_3)_3-N(H)C(O)N(H)-(CH_2)_2-N<[(CH_2)_2-N(H)]_2\}$, domains; specifically by their strongly ligating tertiary amine, $-N<$, and carbonyl, $>C=O$, groups (13,27,28), while their encapsulating ability results from the steric hindrance to the diffusion of complexed species imposed by the highly branched architecture of dendritic domains (a "golf-ball-in-the-bush" effect: imagine trying to retrieve a golf ball that is buried deep in a bush along the side of a fairway). Such compositions result in the pronounced affinity of these networks towards various electrophiles and strong attraction of complexed cations, such as Ag^+, Cu^+, Cu^{2+}, Ni^{2+}, Cd^{2+}, Fe^{2+}, Fe^{3+}, Au^{3+}, Co^{2+}, Pd^{2+}, Pt^{2+}, Pt^{4+}, Eu^{3+}, Tb^{3+}, etc., or organic compounds, such as methylene blue, methyl red, Rose Bengal, etc., as illustrated for Cu^{2+} as an example in Figure 4 (27,28).

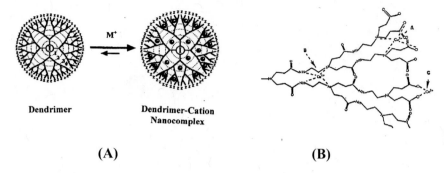

(A)   **(B)**

Figure 4: Encapsulation of electrophilic reagents, such as organometallic antifoulants, inside the nucleophilic dendritic domains (i.e., circular cells of Figures 1 and 3). (A): Absorption of cations (M$^+$) in highly branched and confined dendritic environment. (B): Example of complexation of Cu^{2+} cations inside a PAMAM domain by the tertiary nitrogen branch junctures (-N<) and neighboring amide groups carbonyl >C=O groups (according to Reference 27).

In addition, the resulting coatings are exceptionally smooth (typical roughness is less than 1.5 nm by atomic force microscopy, AFM), and they show surface properties of methylsilicones, including surface energies of about 21-24 mN/m and advancing contact angles of about 110° to water, 70-76° to CH$_2$I$_2$ and 31° to C$_{16}$H$_{34}$, all of which are highly desirable for the targeted non-stick applications.

In summary, these silicon-containing dendritic polymer coatings show the following unique combination of properties that are ideal for antifouling applications:

(a) Surface properties characteristic of organo-silicones, including flexibility, low surface energy, high contact angle(s), wetting characteristics, chemical inertness and non-stick behavior.

(b) Pronounced nano-smoothness that is beneficial in preventing or at least greatly reducing the ability of fouling organisms to physically attach.

(c) Unprecedented ability to encapsulate *and retain* metallic cations, such as functionally effective copper repellants of fouling organisms.

(d) Because the nano-sized cells of these dendritic polymer coatings are distributed in a regularly organized fashion right under the outer surface (see Figure 1), the entrapped repellants are brought in direct contact with incoming fouling organisms, yet tightly bound by the strong chemical complexation forces and severe steric hindrance of the highly branched dendritic structure to the coatings. Thereby, they are effective against the foulants yet prevented from leaching and endangering the surrounding aquatic environment and its inhabitants.

(e) Elimination or at least significant reduction of repellant(s) leaching also results in the reduction of the amounts of repellant(s) needed to ensure adequate antifouling protection for any given period of time.

Therefore, these coatings require significantly lower amounts of antifoluants to match the effectiveness of the traditional antifouling paints (which often contain as much as 50-80 wt% of the latter), and hence create none or significantly less pollution problems with either environmental release or disposal during cleaning and re-coating operations.

Preparation of Nano-Domained Dendritic Polymer Coatings

Two main families of dendritic polymers that have been used for the preparation of nano-structured, honeycomb-like coatings of Figure 1 include the methoxy-functionalized radially-layered copolymeric poly(amidoamine-organosilicon), PAMAMOS, dendrimers and their closely related hyperbranched polyurea-organosilicon, HB-PUSOX, counterparts. Methoxy-functionalized PAMAMOS dendrimers, with trimethoxysilyl, $-Si(OCH_3)_3$, or dimethoxymethylsilyl, $-Si(OCH_3)_2(CH_3)$, organosilicon (OS) end-groups are now commercially available from Dendritech, Inc., Midland, MI (29), and they can be crosslinked into the nano-structured honeycomb-like coatings of Figure 1 by a typical sol-gel chemistry shown in Reaction Scheme 1 (30). In this approach, crosslinking parameters, such as reaction conditions, times and temperatures, can be controlled to yield coatings with controlled mechanical properties. The crosslinking reaction rates can be controlled by appropriate selection of catalyst (e.g., traditional organo-tin compounds or a variety of acids, such as fumaric, acetic or benzoic acid), or by eliminating the catalyst altogether since PAMAM interiors are basic enough to self-catalyze the hydrolysis of methoxysilyl groups in the presence of moisture, or by operating at elevated temperatures which may range to about 100-120°C.

Reaction Scheme 1

Methoxy-functionalized HB-PUSOX are prepared in two steps. First, a bimolecular non-linear polymerization (BMNLP) (23-25) of a diisocyanate (e.g., isophorone diisocyanate, IPDI) and a triamine (e.g., tris(2-aminoethyl)amine,

TREN) is performed so as to obtain an amine-terminated polyurea (HB-PU), followed by the addition of a methoxysilyl isocyanate or glycidyl ether (not shown) to this hyperbranched polymer, as shown in Reaction Scheme 2 (23-25). This polymer is then crosslinked into the coating by the same type of sol-gel chemistry as shown above for PAMAMOS dendrimers.

Reaction Scheme 2

In HB-PUSOX, the R end-groups can be CH_3 or C_2H_5 and theses polymers can be considered as hyperbranched analogues of PAMAMOS dendrimers with hydrophilic and nucleophilic molecular "interiors" (PU) and organosilicon (OS) "exteriors" (23). Also like in the case of their more precise PAMAMOS relatives (20,30), their crosslinking (23-25) can be performed with water vapor (including moisture from the air) or liquid water, under controlled humidity or in open air, in the presence of suitable catalysts to regulate the reaction rate or without them, and to high degrees of crosslinking which directly predetermines mechanical properties and swelling of the resulting networks. In the cases where polyurea is nucleophilic enough, and very fast crosslinking is not of crucial importance, the use of a catalyst may not even be required.

The crosslinked coatings obtained from these dendritic precursor polymers usually have quite smooth surfaces which are often hydrophobic (advancing contact angles of water ranging from 80 to 110°) indicating methylsilicone groups at the surfaces, and can exhibit very high pencil hardnesses ranging to above 7-8 H on hard supports. In addition, depending on the composition of the base polymer, these coatings can serve as very good matrices for a variety of different additives, of which some of the best antifouling performances were obtained with copper and zinc compounds.

Antifouling Formulations

While the synthesis of HB-PUSOX was originally developed as a two-step process with isolation and purification of the intermediate HB-PU (23-25), it was subsequently improved and optimized to a simple one-pot procedure in which the product was obtained quantitatively in methanol (MeOH), or isopropanol (IPA), without intermediate isolation. Of the two solvents, MeOH was found preferable because it provides for an extended shelf-life of the final product. The reason for this is the formation of MeOH as the by-product of the $Si-OCH_3$ end-groups hydrolysis (see Reaction Scheme 1) which precedes the undesired crosslinking reaction on storage (i.e., gelation) via silanol condensation, and the fact that the former reaction is significantly inhibited in MeOH as the reaction environment. More recently, another type of HB-PU was developed in order to achieve more economically desired products. This new polymer was based on diethylene triamine (DETA), hexamethylene diisocyanate (HMDI) and glycidoxypropyl-trimethoxysilane (GTMOS), or different combinations of these reagents and IPDI and isocyanatopropyltrimethoxysilane (ICTMOS). The obtained products have sufficient shelf-life for practical applications, and can be cast in excellent coatings, as judged by their mechanical strength, optical clarity and transparency.

Over 200 different antifouling formulations (see Figure 5) were prepared using the above-described different types of HB-PUSOX and a variety of different biocides, including copper acetate, copper oleate, copper sulfate, Nuocide 1051, benzoic acid (BA), terephthalic acid, tannic acid, zinc and copper pyrithone (aka zinc and copper Omadine, ZnOm and CuOm, respectively), copper thiocyanate, copper oxide, triphenylboron-pyridine (aka Borocide P, BP) and niclosamide. The formulations were prepared in both MeOH and IPA as solvents at polymer concentrations ranging from 10 to 65 wt%, and coated and tested on stainless steel, glass and glass fiber reinforced epoxy plates. These formulations were prepared in five different chronologically progressive "generations". The first generation used tin catalyst for crosslinking, roughly 65 wt% total dissolved solids (TDS) paint concentration and very small amounts (1-5 wt%) of biocides, such as copper oleate, BA, ZnOm, CuOm, BP and tannic acid. It was found that BA acted as a very powerful crosslinking agent, most likely causing extensive siloxane redistribution/equilibration reactions, and hence in the second generation of coatings tin catalyst was replaced with BA, the optimal amount of which was found to be 20 wt%. This finding also suggested that BA might fulfill both roles of biocide and crosslinker, but it was later found by Fourier transform infrared spectroscopy (FTIR) that it was not present in the coating after 30 days of exposure to water. Nevertheless, BA was so effective in reducing the crosslinking reaction time while increasing the degree of crosslinking and the resulting hardness of the coatings that it was selected as the catalyst of choice for generation 3 and 4 coatings in place of the previously used tin catalysts. In addition to this, generation 3 and 4 coatings focused on commercial biocides ZnOm, CuOm and BP. Furthermore, the amounts of biocides were gradually increased from 5 to 40 wt%, with the best results obtained from formulations containing 15-20 wt% of each in a 1:1 wt. ratio.

The obtained coatings were evaluated for their uniformity, thickness, hardness and contact angles to water, and the samples were examined under an optical microscope. All samples were uniform when produced as single-coat brush-on coatings, averaging in thickness from about 50 to about 120 μm, with hardnesses that directly depended on the amount of BA used for crosslinking. For example, the values obtained by pencil hardness tests ranged from as low as 5B for TREN-IPDI coatings with no BA added to over 5H for TREN-IPDI coatings with 20 wt% BA after 12 h of crosslinking, or even 7H for DETA-HMDI coatings with or without BA after the same period of time. In general, DETA-based coatings were significantly harder and more scratch resistant than the corresponding TREN-IPDI coatings, and with 20 wt% BA achieved 5H hardness already after 6 h. Consequently, hardness and scratch resistance became the criteria for judging the robustness and mechanical properties of these coatings, both of which directly impact their long-term usage and durability.

Various copper biocide-containing coatings were also evaluated for their chemical structure by near edge X-ray absorption fine structure (NEXAFS) at the National Synchrotron Light Source at Brookhaven National Laboratory. The bulk of this work was done on PAMAMOS dendrimer-based coatings where it was clearly shown that copper cations preferentially complexed with PAMAM tertiary amines and carbonyl oxygens, as expected, and that only when the PAMAM capacity for complexation (which was generationally dependent) was exceeded the excess copper "spilled over" into the OS domains where the cations complexed with the most nucleophilic component, the siloxy oxygens (29,32). With HB-PUSOX networks, most of the work was done with early generation coatings which unfortunately contained too small amounts of copper for easy detection. Nevertheless, the obtained results indicated that copper cations preferentially interacted with urea carbonyl oxygens, while their interactions with tertiary and urea nitrogens were much weaker than in the closely related PAMAMOS dendrimers. A similar situation was also found in 15 wt% CuOm and ZnOm containing samples which indicates that while PAMAMOS are clearly stronger chelators than HB-PUSOX, the latter can nevertheless successfully complex the cations and keep them bound to the host matrix, preventing their mobility and eventual leaching.

Figure 5: *Several different antifouling paints prepared from PAMAMOS dendrimers and HB-PUSOX. Note that because the base polymer solutions are colorless, different colors originate from different antifoulants used. Small vials on the tops of the respective paint containers contain catalyst solutions that may or may not be needed (see text for explanation).*
(see page 4 of color insert)

PAMAMOS Dendrimers-Based Coatings

PAMAMOS dendrimer-based antifouling formulations were fist used in a model study to obtain the proof of concept before evaluating the closely related hyperbranched polymer formulations which were deemed economically more feasible. Various formulations described in the preceding section were painted on one side of 3"x5"steel plates while the other side was left unpainted for comparison. The plates were tested for their antifouling behavior in the highly zebra mussel-infested water of Sanford Lake, MI and small (1"x3") painted glass plates were exposed in laboratory containers to the same water for leaching studies (see Figure 6). Samples of water were taken from these containers periodically and evaluated by an independent analytical laboratory for the content of biocide used in every particular formulation. Determinations were performed by atomic absorption spectroscopy (AAS) which had the detection limit of 0.4 ppm for the elements tested, and in all cases no contamination was detected even after a year long exposure. These results proved the concept that PAMAMOS dendrimer-based nano-domained coatings can be formulated for environmentally benign antifouling coatings and they encouraged further investigations with the closely related hyperbranched polymer systems.

A   B

Figure 6: *Testing of PAMAMOS-based antifouling paints. (A) Sample preparation: steel plates (lower left corner) ready for testing in Sanford Lake, MI; vials with samples submerged in the same lake water (upper center and right) for leaching tests. (B) Test plates from A after 3 months in Sanford Lake (according to Reference 32): notice small specks on the lower row of plates (unpainted sides) representing settled zebra mussels and the settlement-free upper row of plates (painted sides).*

Seawater Studies

Two different types of seawater studies were performed predominantly with hyperbranched polymer formulations. In *dynamic tests,* three different sets of 10 patches (5 on the port and 5 on the starboard sides) were coated on the hulls of three seafaring vessels (two in the Atlantic and one in the Pacific) operated by the Sea Education Association of Woods Hole, MA, and their performance was monitored at specified time intervals (mostly monthly) by divers. In *static tests,* two different barges of 21 and 24 samples, respectively, were submerged off the docks at two different locations in the Gulf of Mexico, on the coasts of Florida and Alabama, respectively.

Dynamic Studies: First, a set of five generation 1 HB-PUSOX coatings was tested from September 2005 until May 2006 on the *SSV Corwith Cramer* which sailed over 6,000 miles from Rockland Marine, MA, to the Caribbean and back, with stops in Key West (Florida), Carriacou (Grenada), Grenadines, and Roatan (Honduras), among other places. For the first three months of the trip, the coating performance was satisfactory; but after that time, significant biofouling was observed (see Figure 7). The most prominent foulants were pelagic gooseneck barnacles (*Lepus ansifera*), small striped barnacles (*Balanus amphitrite*), striped gooseneck barnacles (*Conchoderma virgatum*), spiral tube worms (*Hydroidus dianthus*), green seaweed (*Spongomorpha*) and transient isopods, all consistent with the warm areas of the Atlantic Ocean.

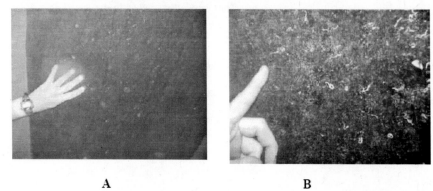

A   B

Figure 7: *Underwater photographs of selected test patches on the hull of SSV Corwith Cramer after (A) 1 month and (B) 4 months on its trip through warm waters of the Atlantic Ocean.*

Following this, in March 2006, a set of generation 2 coatings was tested on the hull of the *SV Captiva* which sailed during the summer of 2006 in the coastal waters of Massachusetts. The hull was inspected under water twice, in July and in August, and the vessel was hauled out of the water in October. On the first inspection all test squares had a light film deposit identical to that on the commercial E-paint used on the rest of the hull, and there was significant growth on the starboard test squares but none on the portside squares. On the second inspection, there was some initial growth on the portside squares, and heavy growth on the starboard squares, probably because of the different exposure to sunlight when the vessel was in dock.

Finally, in June of 2006, a set of generation 3 formulations was applied to patches on the hull of the *SSV Robert C. Seamans,* which spent the fall and winter of 2006-2007 in the eastern and central Pacific Ocean. The voyage started from Lake Union, Seattle, WA, to San Diego in mid-October, through

Puerto Vallarta (Mexico), Marquesas, Papeete (Tahiti), again Marquesas, to Hawaii, crossing the equator with stops at Palmyra Atoll and the Line Islands. Diver inspections were performed from November 2006 to April 2007 on a monthly basis. After the 6-week trip from Lake Union to San Diego, the growth consisted mainly of small clumps of brown algae and less than 1 cm large gooseneck barnacles. After 3 months, however, the growth amount increased and all patches were about 90% covered in thick algae growth and filamentous diatoms. However, when the ship reached La Paz, Mexico, the main types of growth were worm casings, hydroids, gooseneck barnacle larva, several types of green and brown algae and bryozoan colonies. Interestingly, however, the average thickness of the growth layers was less than that seen 4 weeks earlier in San Diego, indicating that some of the growth was rather loose and had been washed off during the travel.

During the next 4 months a noticeable growth accumulated on all patches, consisting primarily of small barnacles (65%), bryozoans (5%), algae (7%), and some occasional white hydroids (3%). Some barnacles were over 1 cm in diameter and height. On the return trip, the entire bottom, including the test patches was cleaned by divers in Honolulu harbor in accordance with the agreement with the Nature Conservancy in Palmyra Atoll Lagoon. Hence this became effectively a brand new growth cycle. By the end of the voyage in April, 2007, the test patches were again over 60% covered, primarily with encrusting algae, a few small barnacles, and small tube worm casings.

Static Studies: These studies were performed on floating barges with generation 4 and 5 coatings. Barge # 5 had 12 coating samples on each side, and was placed in Palm Harbor, Florida, in November 2006 (see Figure 8). Included for comparison were two samples from the best performing commercial paints: *Pettit Vivid* (good for both hard and soft growth prevention), and *Pettit Trinidad* (good for hard fouling prevention), and also a waxed polyester gelcoat. Our formulations were based on TREN-derived HB-PUSOX and it was noted that after 3 months of exposure some of them, particularly those containing between 15 and 25 wt% ZnOm and BP (see H-J of Figure 8), equaled if not bettered the commercial standards, representing a major breakthrough in the antifouling ability of our coatings.

Barge # 6 had 16 coating formulations on each side, and it was placed in Gulf Shores, Alabama, in March 2007. It contained several commercial coatings for comparison, including *Pettit Vivid*, *Pettit Trinidad*, *E-Paint ZO*, *Interlux VC-17m*, *Interlux Micron CSC* and *West Marine Bottomshield*, a waxed polyester gelcoat, as a control, and for the first time samples based on DETA-derived formulations with 15 wt% ZnOm-BP combination. It was found that a number of our samples performed as well as or better than the commercial products (particularly those containing both ZnOm and BP), but the best performance was obtained from the samples containing 25 wt% ZnOm and 25 wt% BP.

180

A   B

Figure 8: Generation 4 coatings on one side of Barge # 5 tested in Palm Harbor, FL after the (A) 1 month and (B) 3 months exposure. Samples A-D were coated with commercial and control paints. Notice exceptionally good performance of samples H-J from our coating formulations.

Freshwater Studies and Zebra Mussel Health Studies

Freshwater studies were performed in Coldwater Lake, Saginaw Bay, Sanford Lake (all MI) and off the coast of Beaver Island in Lake Michigan. The formulations tested ranged from generation 1 through generation 4 and they were exposed from 3 months (late generations) to over 1.5 years (early generations). The Coldwater Lake and Beaver Island sites were extensively controlled and observations on some of the newer formulations are still in progress. Plate grid protocols and grid design were developed and examined (see Figure 9).

The Beaver Island location was predominantly infested with quagga mussels, and did not have any significant recruitment from May 2006 to June 2007, with only an occasional mussel being observed on the concrete bricks or PVC piping used to construct the plate grids. In contrast, the plate grids in Coldwater Lake showed a high density of zebra mussels, which attached over the approximately one-year period. New plates (as the new generations of formulations were developed) were added during the year and the grids were monitored throughout. Settlement, due to metamorphosis, was first evident in late July, 2006, and continued over the following months, except from December to March. The relatively clear plates had a significant accumulation of mussels at the top of each plate. This was caused by mussel mobility, which enabled immigration of mussels from the grid PVC piping. The higher loaded biocide formulations (i.e., later generations) showed significantly reduced mussel deposition (see Figure 9B).

A   B

Figure 9: Freshwater studies. *A:* A typical sample grid used in Lake Michigan studies. *B:* A similar sample grid from the Coldwater Lake study (according to Reference 23).

In laboratory tests, the later generation formulations also showed quite promising results. These formulations effectively stressed the mussels, yet biocide leaching was not significant. Test plates with formulations coated on top of Ameron 235 primer (the same substrate used on the *Corwith Cramer* (see sea water studies described above) and not expected to have any antifouling activity) were compared with a control having only Ameron prime coat. The plates were placed in aquaria containing lake water; and sixty zebra and quagga mussels collected from St. James Harbor, Beaver Island, MI, were introduced per test, thirty of them on each plate and thirty at the far end of the aquarium (see Figure 10). All mussels were larger than 30 mm, meaning that they were about three years old. The mussels were considered settled on a surface if they were orientated dorsal side up and had laid down a byssal thread. The number of mussels that moved off the plate was monitored as a function of time and their health was evaluated by determining the chloride, sodium, potassium and calcium ion levels in their hemolymph. When mussels experience sub-lethal stress, their hemolymph calcium levels increase while sodium and chloride levels decrease. As expected, the Ameron 235 control surface and other untreated surfaces within the aquaria did not affect the mussels in any way, nor did they result in elevated calcium levels or any other adverse effect on their health.

A low mortality rate was observed in the control and test plate mussels except for the coatings with 25 wt% ZnOm (the highest ZnOm concentration tested) which did show mortality in both the test plate and control mussels. The mussel mobility was high in 15 wt% ZnOm/5 wt% BP, 15 wt% ZnOm/15 wt% BP, 25 wt% CuOm, and 25 wt% copper thioisocyanate/2.5 wt% ZnOm compared to control, which indicated an attempt by the mussel to avoid noxious areas. Very little movement was observed with the commercial coat (i.e., Petite Vivid) and the 25 wt% ZnOm formulation, which implied a more toxic environment. The circulating hemolymph concentrations of chloride, sodium and calcium confirmed these observations. Mussels that showed stress, based on a decrease in hemolymph chloride and sodium with a concomitant increase in hemolymph calcium, were those that were exposed for one week or more to test plates containing 15 wt% ZmOm (or higher) or 15 wt% BP formulations. Similar results were observed with quagga mussels, although mortality was higher probably because of the species difference. Again, test plate formulations with 15 wt% (or higher) ZnOm and BP content significantly affected mussel attachment and caused a decrease in their hemolymph chloride and sodium.

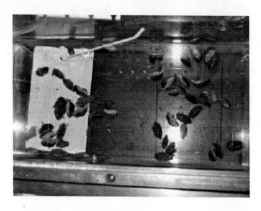

Figure 10: *Aquarium test plates. A typical plate (left) and control (uncoated plate: right) during the test. Migration of mussels off the plates was followed.*

Leaching Studies

Over 200 different water samples were exposed to over 20 different hyperbranched polymer coating formulations in tests that lasted from 1 hour to 7 months. The coatings were prepared on primed steel panels and left to stand in jars filled with a fixed volume of Sanford Lake water at room temperature. Depending on the nature of the biocide, the samples were analyzed in an overwhelming majority of cases for copper or zinc by AAS which had a detection limit of 0.3 ppm, and for benzoic acid by HPLC with a detection limit of 10-100 ppb. For early generation coatings, the biocide level in water was below the detection limit of AAS, while in comparison, zinc leached out of E-Paint ZO in 0.6, 1.9, 1.0 and 0.6 ppm after 1, 2, 3 and 4 months, respectively, or from Petit Vivid in 0.5, 0.6 and 0.9 ppm, after 1, 3 and 5 weeks, respectively. Even the later generation coatings, containing 15-40 wt% ZnOm, showed significantly (i.e., by one to two orders of magnitude) less leaching than the corresponding commercial paints, as shown in Table 1.

In addition to this, the ongoing dynamic evaluation of generation 5 coatings indicates that leaching of organic pyrithione ions, $^-SC_5H_4NO^-$, is also at least ten times less than from any other previously tested commercial coating. All these data clearly prove the environmental benignity of these novel nano-domained coatings.

Table 1: Comparison of leaching data from selected coating formulations

Polymer	Biocide	Initial biocide content, wt%	Duration of test, months				
			0.25	1	3	5	7
Petit Vivid	ZnOm/ Cu-TI	25/2.5	1	2	3	3	5
PAMAMOS	CuAc	2.5	0.01	nd	nd	0.03	nd
HB-PUSOX	CuAc	4	0.01	nd	0.01	nd	nd
HB-PUSOX	ZnOm/ Cu-TI	25/2.5	0.02	0.1	0.1	0.1	0.1
HB-PUSOX	ZnOm/BP	15/15	0.3	0.3	0.3	0.4	0.2
PAMAMOS/HB-PUSOX (1:1)	ZnOm/BP	15/15	0.2	0.2	0.3	0.4	0.3

Concentrations of leached biocides are normalized with respect to Petit Vivid performance after 1 week which was 0.4 ppm. TI – thioisocyanate. HB-PUSOX data for TREN/IPDI systems; PAMAMOS data for generation 3 dendrimer. Nd = not detected.

Conclusions

A new type of antifouling coatings has been developed using dendritic polymers (i.e., dendrimers and hyperbranched polymers) as building blocks and a controlled, "bottom-up" thin-film construction strategy. These coatings have low energy silicone-like surfaces and honeycomb-like 3D structures in which active biocides are immobilized by both chemical complexation and steric encapsulation inside nanoscopic dendritic cells which permits their unchanged antifolung activity while preventing their leaching into the outside aquatic environment. As a consequence these coatings for the first time combine all of the benefits of the two main classes of traditional antifouling coatings (i.e., low surface energy silicones and biocide-containing ablative ones) with an added feature of providing complete or significantly enhanced environmental benignity.

Over 50 different formulations of these novel coatings from two different types of dendrimers, three different types of hyperbranched polymers, and over ten different biocides were prepared and evaluated in this work. Their antifouling efficacy was tested in real life situations on hulls of three seafaring boats, on over ten stationary rigs and barges in fresh and sea waters, and in aquarium laboratory studies for their effects on the health of zebra and quagga mussels. Their environmental benignancy was tested in deionized, as well as in real lake water, and it was found that at comparable or better antifouling activities, there was either no detectable leaching of biocides, or if it occurred it was at least one to two orders of magnitude less than that from the best available commercial antifouling paints. It should be also noted that these new antifouling coatings require drastically reduced amounts of biocides for adequate antifouling activity.

Acknowledgements

The authors gratefully acknowledge funding of this work by two grants from the National Science Foundation: contract numbers DMI-0419193 and DMI-0522183. Samples of copper and zinc Omadine were graciously supplied by Arch Chemicals.

References

1. Reisch, M., *Chem. Eng. News*, **2001**, *November. 12*, 14.
2. Stebbing, A., *Marine Polution Bulletin*, **1985**, *16*, 383.
3. Kannan, K.; Senthilkumar, K.; Loganthan, B.; Takahasi, S.; Odell, D.; Tanabe, S., *Environ. Sci. Tech.*, **1997**, *31*, 296.
4. Stein, J.; Truby, K.; Wood, C.D.; Wiebe, D.; Montemarano, J.; Holm, E.; Wendt, D.; Smith, C.; Meyer, A.; Swain, G.; Kavanagh, C.; Kovach, B.; Lapota, D., *Polym. Prepr*, **2001**, *41(1)*, 236.
5. Burnell, T.; Carpenter, J.; Truby, K.; Serth-Guzzo, J.; Stein, J.; Wiebe, D., *"Advances in Non-Toxic Silicone Biofouling Release Coatings"*, in *"Silicones and Silicone-Modified Materials"*, Clarson, S.J.; Fitzgerald, J.J.; Owen, M.J.; Smith, S.D., Eds., ACS Symposium Series 729, Am. Chem. Soc., **2000**, pp 180-193.
6. Lindner, E. in *Recent Developments in Biofouling Control*, Thompson, M.-F., Ed., Balkema, Rotterdam, **1994**, pp 305-311.
7. Lindner, E., *Biofouling*, **1992**, *6(2)*, 193.
8. Dvornic, P.R., *"Thermal Properties of Polysiloxanes"*, in *"Silicon-Containing Polymers"*, Jones, R.G.; Ando, W.; Chojnowski, J., Eds., Kluwer Academic Publishers, Dordrecht, **2000**, pp 185-212.
9. Owen, M.J., *"Surface Properties and Applications"*, in *"Silicon-Containing Polymers"*, Jones, R.G.; Ando, W.; Chojnowski, J., Eds., Kluwer Academic Publishers, Dordrecht, **2000**, pp 213-231.
10. Baier, R.E.; Meyer, A.E., *Biofouling*, **1991**, *5*, 349.
11. Dvornic, P.R.; de Leuze-Jallouli, A.M.; Owen, M.J.; Perz, S.V., *"Radially Layered Poly(amidoamine-organosilicon) Copolymeric*

Dendrimers and Their Networks Containing Controlled Hydrophilic and Hydrophobic Nanoscopic Domains" in *"Silicones and Silicone-Modified Materials"*, Clarson, S.J.; Fitzgerald, J.J.; Owen, M.J.; Smith, S.D., Eds., ACS Symposium Series 729, Am. Chem. Soc., **2000**, pp 241-269.

12. Dvornic, P.R.; de Leuze-Jallouli, A.M.; Owen, M.J.; Perz, S.V., *U.S. Patent*, 5,902,863, **1999**.

13. Balogh, L.; de Leuze-Jallouli, A.M.; Dvornic, P.R.; Owen, M.J.; Perz, S.V.; Spindler, R., *U.S. Patent*, 5,938,934, **1999**.

14. Dvornic, P.R.; Lenz, R.W., *"High Temperature Siloxane Elastomers"*, Hüthing & Wepf, Heidelberg, **1990**.

15. Dvornic, P.R.; Tomalia, D.A., *Sci. Spectra*, **1996**, *5*, 36.

16. Tomalia, D.A.; Dvornic, P.R., *"Dendritic Polymers, Divergent Synthesis (Starburst Polyamidoamine Dendrimers)"*, in Salamone, J.C.; Ed., *"Polymeric Materials Encyclopedia"*. CRC Press, Boca Raton, Vol. 3: 1814-1830, **1996**.

17. See, for example: (a) Newkome, G.R.; Moorefield, C.N.; Vögtle, F.; *"Dendrimers and Dendrons. Concepts, Synthesis, Applications"*; Wiley-VCH Verlag, Weinheim, Germany, **2001**. (b) Fréchet, J.M.J.; Tomalia, D.A., Eds., *"Dendrimers and Other Dendritic Polymers"*, J. Wiley and Sons, Chichester, UK, **2001**.

18. See, for example: (a) Mishra, MK; Kobayashi, S, *"Star and Hyperbranched Polymers"* ,Marcel Dekker, New York, **1999**. (b) Sunder, A; Heinemann, J; Frey, H, *Chem. Eur. J.,* **2000**, *6(14)*, 2499. (c) Hult, A, *"Hyperbranched Polymers"*, in Mark, HF, Ed., *"Encyclopedia of Polymer Science and Technology"*, 3rd ed. Wiley-Interscience, New York, **2003**, Vol 2, pp. 722-743. (d) Gao, C; Yan, D, *Prog. Polym. Sci.*, **2004**, *29(3),* 183.

19. Dvornic, P.R.; Owen, M.J., Eds., *"Silicon-Containing Dendritic Polymers"*, Springer, Guildford, UK, **2009**.

20. Dvornic P.R.: Owen, M.J., *"Silicon-Organic Dendrimers"*, Chapter 11 in Reference 20, pp 285-314.

21. Dvornic, P.R.; de Leuze-Jallouli, A.M.; Owen, M.J.; Perz, S.V., *Macromolecules*, **2000**, *33*, 5366.

22. Dvornic, P.R.; de Leuze-Jallouli, A.M.; Owen, M.J.; Perz, S.V., *U.S. Patent*, 5,739,218, **1998**.

23. Dvornic, P.R.; Meier, D.J., *"Hyperbranched Silicon-Containing Polymers via Bimolecular Non-Linear Polymerization"*, Chapter 16 in Reference 20, pp 401-419.

24. Dvornic, P.R.; Hu, J.; Meier, D.J.; Nowak, R.M.; Parham, P.L., *U.S. Patent, 6,534,600 B2*, **2003**.

25. Dvornic, P.R.; Hu, J.; Meier, D.J.; Nowak, R.M., *U.S. Patent*, 6,384,172 B1, **2002**.

26. Dvornic, P.R.; Li, J.; de Leuze-Jallouli, A.M.; Reeves, S.D.; Owen, M.J., *Macromolecules*, **2002**, *35*, 9323.

27. Ottaviani, M.F.; Montali, F.; Turro, N.J.; Tomalia, D.A., *J. Phys. Chem., Chem. B.*, **1997**, *101(21)*, 158.

28. Bubeck, R.A.; Dvornic, P.R.; Hu, J.; Hexemer, A.; Li, X.; Keinath, S.E.; Fischer, D.A., *Macromol. Chem. Phys.*, **2005**, *206*, 1146.
29. www.dendritech.com
30. Dvornic, P.R.; Li, J.; de Leuze-Jallouli, A.M.; Reeves, S.D.; Owen, M.J., *Macromolecules*, **2002**, *35*, 9323.
31. Dvornic, P.R.; Bubeck, R.A.; Reeves, S.D.; Li, J.; Hoffman, L.W., *Silicon Chem.*, **2003**, *2*, 207.
32. Dvornic, PR; *J. Polym. Sci. Part A: Polym. Chem.*, **2006**, *44*, 2755.

Chapter 12

Surface Modification of Polyamide Reverse Osmosis Membranes with Hydrophilic Dendritic Polymers

Abhijit Sarkar*, Peter I. Carver, Tracy Zhang, Adrian Merrington, Keneth J. Bruza, Joseph L. Rousseau, Steven E. Keinath and Petar R. Dvornic

Michigan Molecular Institute, 1910 West St. Andrews Road, Midland, MI 48640-2696; Phone: +1-989-832-5555; email: sarkar@mmi.org

The effects of surface modification of commercial reverse osmosis (RO) polyamide membranes were studied as a function of flux/rejection and surface properties. Selected ultra-low pressure RO membranes were modified by *in situ* crosslinking of either hyperbranched polymers (HBPs) or polyamidoamine (PAMAM) dendrimers and polyethylene glycols (PEG). As expected, surface and chemical modification of membranes significantly reduced their contact angles as measured by dynamic as well as static contact angle measurement techniques. Lower contact angles indicate more hydrophilic membranes with potential for increased resistance to fouling by hydrophobic foulants, such as biofoulants and organic materials.

During the past two decades, the RO process has gained extensive attention in reclamation of water and separation of organics from aqueous streams (*1-3*). However, successful utilization of RO technology is often hampered by fouling, which poses a major obstacle for the membrane application (*4-6*). Therefore, there is a growing interest in surface modification of existing RO membranes to introduce properties that markedly reduce fouling, specifically biofouling, while retaining desirable permeate water flux and rejection characteristics (*7-10*).

Commercial RO membranes are typically made of thin-film-composite aromatic polyamides for which the microorganism-induced biofilm formation is a major problem. This biofouling occurs on the active membrane surface and it increases the operating pressure by about 50%, requiring regular cleaning by chlorine treatment.

Table I. Selected specifications for commercially available RO membranes.

Membrane	HR98PP	CA995PE	SEPA-MS05	SEPA-SS1C	DESAL-3B	Elix® 3 RO Cartridge
Manufacturer	Dow/ FilmTec	Dow/ FilmTec	GE-Osmonics	GE-Osmonics	Desalination Systems, Inc.	Millipore
Composition	Polyamide thin film composite	Diacetate/ polyester	Polyamide	Cellulose acetate	Polyether-sulfone	Polyamide thin film composite
Typical NaCl rejection (%)	>97.5	>95	>98	>98	98.5	94-99
Chlorine tolerance	Low	Low	Low	Low	Low	Low
pH range	2-11	2-8.5	3-11	2-8	4-11	4-11
Tendency for biofouling	High	Low	High	Low	High	High
Maximum operating pressure (psi)	870	1015	1015	1015	650	65
Maximum operating temp (°C)	60	35	80	50	50	35

Table I summarizes the selected of representative commercially available RO membranes and shows that all polyamide RO membranes are prone to biofouling. In fact, apart from the cost of energy to run the high-pressure pumps, membrane fouling is the single most important factor that controls the cost of RO water purification unit (ROWPU) operations. It leads to an undesirable reduction of the flux of permeate water by the source water contaminants during the course of operation and requires cleaning every 4 to 8 weeks and changing of membrane elements approximately every 3 years. Since the maintenance and remediation expenses represent ca. 30% of the total operating cost, a new generation of membranes with inherent anti-fouling capability is urgently needed (11).

The research goal of this study was to develop antifouling modifications for commercially available RO membranes by using dendritic nanotechnology. The modifications are based on the experimental evidence reported in the literature that the fouling resistance of a composite polyamide RO membrane can be significantly increased by treating the said membrane with a hydrophilic coating (12,13). For this, low energy aromatic polyamide RO membranes manufactured by FilmTec Corporation were used (14). The syntheses of network polymers and coating procedures were designed so as to achieve enhanced hydrophilicity and surfaces without pinhole defects (*15,16*).

Polyamidoamine (PAMAM) Dendrimers

Dendrimers are a unique class of polymers which play an important role in the emerging field of nano-technology (*17,18*). They are three-dimensional polymers that structurally represent the most regular members of the dendritic class of macromolecular architecture. Their compact tree-like molecular structure provides a rich source of surface functionality that makes them useful as building blocks and carrier molecules at the nanometer level. They are composed of two or more dendrons emanating from a central core which can either be an atom or an atomic group. They are built of branch cells, which are of three main types: a core (which represents a central juncture), interior cells and surface cells with chemically reactive or inert terminal surface groups (Figure 3).

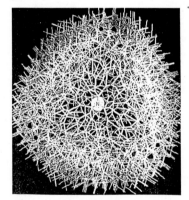 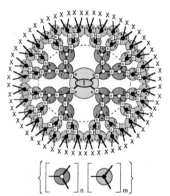

Figure 1. Schematic representation of dendrimer molecular architecture. (Left) a dense-shell model; (right) the branch cell organization.

The structure of these branch cells is determined by the nature of the contributing atoms, bond lengths and angles, directionality, conformational bond flexibility, etc. They can be either homogeneous or differ in their chemical structure, but each of them contains a single branch juncture. Reactive surface groups may be used for continuation of dendritic growth or for modification of reactivity of the dendrimer surface.

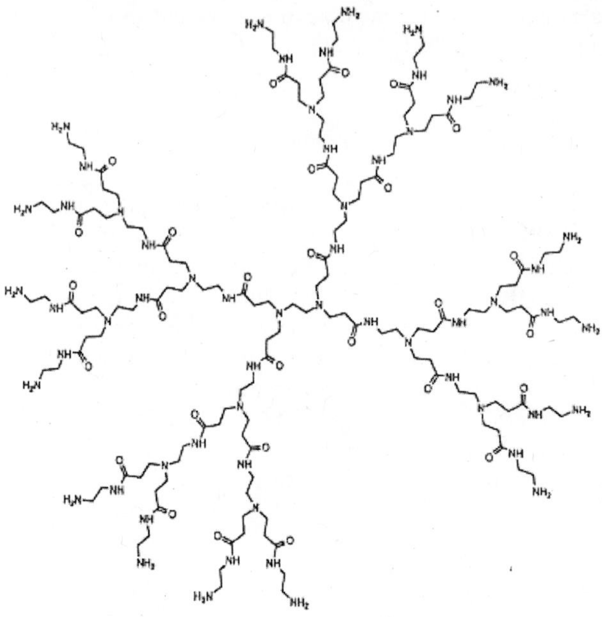

Figure 2. Molecular structure of Generation 2 PAMAM (PAMAM G2.0). At this generation 16 amine groups are exo-presented.

The process of dendrimer formation permits an exceptional degree of reaction control, which results in a very high degree of structural regularity. As a consequence, dendrimers show properties which are not typical for other types of polymers: almost perfect isomolecularity (routinely better than $M_w/M_n = 1.2$ even at very high generations), very well defined molecular size (increasing with generation number in regular increments of ~1 nm/generation), regular molecular shape, and unusually high chemical functionality, due to a large number of surface groups per molecule (*19*). A Generation 2 PAMAM (PAMAM G2.0) dendrimer (Figure 2) was used for this work. The molecular weight of this dendrimer is ~3256 while its diameter is ~29 A. There are 16 exo-presented and terminal functional amine groups in its molecular structure.

Hyperbranched Polymers

Hyperbranched polymers (HBPs) are highly branched, tree-like molecules schematically presented in Figure 3 (*20-23*). Among others, their architecturally related, characteristic properties include: (a) nanoscopic molecular sizes (that range from 1 nm to about 10 nm in diameter and make these polymers ideal building blocks for synthetic film nanotechnology); (b) very high density of molecular functionality (i.e., tens or hundreds of end-groups per molecule that can be either reactive or inert); (c) ability to encapsulate smaller molecular weight species within their highly branched nanoscopic molecular interiors; and (d) significantly lower viscosities than those of linear polymers of comparable molecular weights (hence, much easier and less energy consuming-process). They can be made by several different synthetic strategies of which the most versatile and economical is the so-called bimolecular non-linear polymerization (BMNLP) that has been successfully used for the preparation of a wide variety of different HBP compositions *(24)*. Of particular interest are cross-linked hyperbranched networks (Figure 4) in which HBPs of different chemical compositions can be combined to tailor-make honeycomb-like nano-domained 3D architectures. For example, in such networks, nucleophilic polyamide (PA) or polyurethane (PU) HB nano-domains (a), containing functional groups (c) such as antimicrobial (AMB) or charged groups (CG), can be embedded in a hydrophilic matrix, such as poly(ethylene glycol) (PEG) (b).

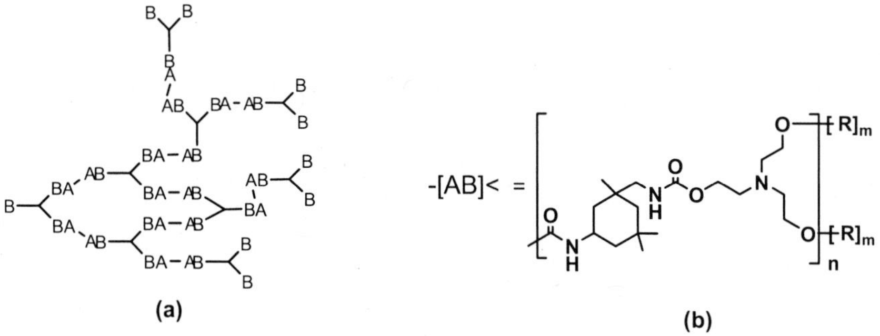

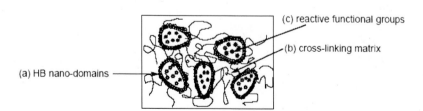

(a)   (b)

Figure 3. Unique molecular architecture of (a) HBP, and (b) representative example of HB polyurethane (HB-PU) with end-groups, R. Note that as a consequence of their structure, the HBPs of (b) can have a hydrophilic molecular interior, –[AB]<, and a hydrophobic "surface" character, -[R]$_m$, or vice versa.

(c) reactive functional groups

(b) cross-linking matrix

(a) HB nano-domains

Figure 4. Schematic representation of a cross-section of a proposed HBP membrane. Note its honeycomb-like structure composed of egg-shaped HB nano-domains (a) containing reactive functional groups (c) that can be used for attachment of antifouling units embedded in a cross-linked matrix (b).

Until now, very few attempts at utilizing unique features of dendritic polymers, including both dendrimers and HBPs, in cross-flow membrane separation processes have been reported (25). This may be due to the fact that compositionally attractive materials were not available until the recent development of our dendrimer and BMNLP synthesis technologies. Now, with these highly versatile synthetic methodologies at hand, the utilization of some of the unprecedented features of these polymers may exert significant beneficial effects on the performance of RO membranes.

Towards this end, we describe here some recent attempts to utilize this new technology in the coating of polyamide RO membranes in order to increase their hydrophilicity without affecting the separation properties. The coatings were designed to be both static (simple dendritic polymers network) and dynamic (dendritic polymers networks with hydrophilic groups, i.e., polymer brushes), and the resulting increase in hydrophilicity was expected to translate into the corresponding improvement of antifouling.

Experimental

General

^{1}H-NMR spectra were obtained on a Varian 400 MHz NMR spectrometer. Samples were dissolved in $CDCl_3$ with tetramethylsilane as the internal reference standard unless otherwise stated. Infrared spectra were obtained on a Nicolet 20 DXB FTIR spectrometer either as a neat liquid on polished KBr plates or, if the sample was a solid, as KBr pellets. SEC was run on a Waters size exclusion chromatograph with narrow molecular weight distribution polystyrene standards. Eluents were either toluene or chloroform and the column was a PLgel 34-2/34-5. SEMs were recorded on an Amray scanning electron microscope with an energy dispersive X-ray detector used as required.

All commercial starting chemicals were used as received from the supplier unless otherwise specified. All commercial solvents were at a minimum ACS reagent grade and used as received. PAMAM G2.0 was purchased from Dendritech, Inc. and used as received. The polyamide based RO membranes (LE and XLE) were obtained from FilmTec, Inc.

Preparation of Hyperbranched Polyamide (HB-PA) from Tris-(aminoethyl) amine and Dimethyl Succinate

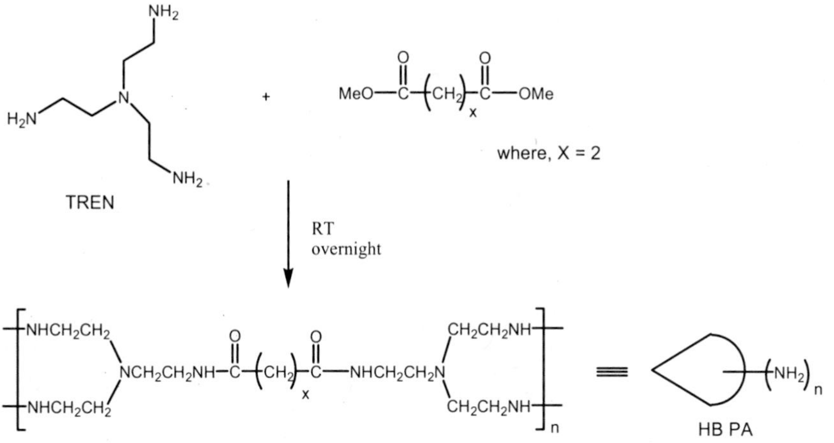

Scheme 1. Synthesis of HB-PA using the BMNLP approach.

Scheme 1 shows the preparation of HB-PA. A 500 mL 3-neck round bottom flask was fitted with an overhead stirrer, a Barret trap and a condenser. Tris-(aminoethyl)amine (TREN) (103 g, 0.704 mol) was added to the flask followed by dimethyl succinate (77.1 g, 0.514 mol). The reaction flask was heated overnight at 110°C in an oil bath. Upon cooling, the viscous yellow product was dissolved in 100 mL of chloroform and precipitated by addition of diethyl ether (2 x 750 mL). The ether layer was discarded and all the volatiles were removed under reduced pressure. Yield : 152 g (84%).

Synthesis of N-hydroxysuccinimide Derivative of Methoxypolyethyleneglycol (MPEG) Acetic Acid

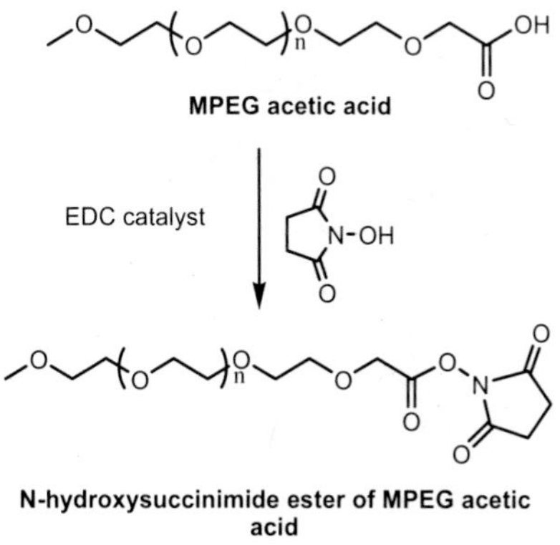

Scheme 2. Synthesis of N-hydroxysuccinimide derivative of MPEG acetic acid.

In a 250 mL round bottom flask equipped with a nitrogen inlet and outlet and a dropping funnel, 1.31 g (6.83 mmol) of 1-(3-dimethylaminopropyl)-3-ethylcarbodiimide hydrochloride (EDC) and 0.780 g (6.78 mmol) of N-hydroxysuccinimide were added and the flask was placed on a magnetic stirrer. Methylene chloride (120 mL) was added to dissolve the reagents in the flask. 4.09 g (6.739 mmol) of methoxypolyethyleneglycol (MPEG) acetic acid (Scheme 2) was added to the stirred solution, drop-wise over 6 minutes. After stirring at room temperature for 42 hours, the reaction solution was transferred to a 1-neck flask and the solvent was removed on a rotavap. The residue was dissolved in water, and extracted 4 times with 200 mL of chloroform each time. The resultant chloroform extract was washed with 500 mL of saturated aqueous sodium bicarbonate solution, followed by 2 x 250 mL of DI water. The organic extract was dried over anhydrous sodium sulfate, filtered, and the solvent stripped under reduced pressure on a rotavap to obtain 1.316 g (yield 49%) of a yellowish solid product. The product was characterized by ^{1}H NMR. ^{1}H NMR (CDCl$_3$): δ 3.50 (s, PEG backbone), 3.21 (s, –OCH$_3$), 2.81 (s, succinimide, 4H), 2.63 (m, CH$_2$CH$_2$, 4H).

Synthesis of MPEG-PAMAM G2.0

1.316 g of the N-hydroxysuccinimide derivative of MPEG acetic acid was placed in a 100 mL round bottom flask and dissolved in 25 mL of methylene chloride. A methanolic solution of PAMAM G2.0 (0.7608 g in 3.55 g methanol, 0.2337 mmol, 3.74 mmol equivalent NH$_2$) was slowly pipetted into the reaction mixture with stirring (Scheme 3). This mixture was stirred at room temperature for 48 hours to allow the MPEG derivative to react stoichiometrically with half of the surface amine groups. The solvent was then removed under reduced pressure and the product subjected to ultrafiltration with a 1000 Dalton nominal molecular weight cut off (MWCO) regenerated cellulose membrane in methanol. The solution after ultrafiltration was stripped under reduced pressure on a rotavap to yield 1.2793 g (yield 64%) of a caramel colored solid product.

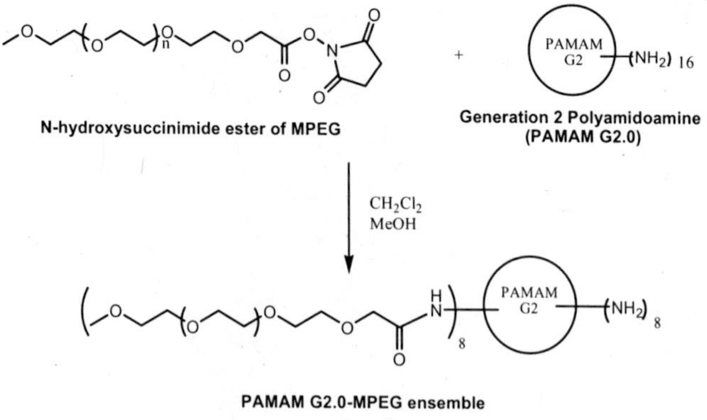

N-hydroxysuccinimide ester of MPEG

Generation 2 Polyamidoamine
(PAMAM G2.0)

PAMAM G2.0-MPEG ensemble

Scheme 3. Synthesis of MPEG-PAMAM G2.0.

The product was characterized by FTIR and NMR. ^{1}H NMR (CDCl$_3$): δ 7.95 (s, CONHCH$_2$CH$_2$N=), 6.41 (d, NHCONH-), 5.91 (d, NHCONH-), 4.30 (d, CHCH), 4.15 (d, CHCH-), 3.50 (s, PEG backbone), 3.21 (s. OCH$_3$), 2.63 (m, CH$_2$CH$_2$, 4H), 2.51-2.46 (m, CONHCH$_2$CH$_2$N=), 1,61 (b), 1.43 (b), 1.23 (s) (CH$_3$); FTIR (KBr): 3263 cm^{-1} (v, NH), 1650 cm-1 (v, C=O), 1547 cm-1 (v, CNH of amide).

General Procedure for Coating the Membrane with HBP-PEG Network

A rectangular section of approximately 19 cm x 23 cm of a FilmTec RO membrane was cut from continuous roll stock and immersed in a deionized (DI) water bath for five minutes to remove the protective glycerin coating. It was then placed in a second bath of DI water for subsequent cleaning. Following this, it was affixed to a glass plate of similar dimensions via masking tape at the four corners making sure that it was flat against the plate. DI water was sprayed between the membrane and the glass plate to ensure that the uncoated polyester "bottom" side of the membrane was always in contact with water. The exposed membrane's polyamide "top" surface was also washed with DI water. Coating solutions were prepared with varying weight percents of HB-PA and stoichiometric equivalents of polyethyleneglycol diglycidyl ether with m.w. 526 (PEG-DGE 526) in water. The solution was poured onto the membrane to cover the entire surface and allowed to crosslink for 30 seconds in a stationary horizontal position, and then an aluminum roller was rolled over the membrane twice with little to no vertical pressure to remove excess solution. A pinhole-free coating coverage was determined by using a dye solution and UV detection (*vide infra*).

General Procedure for Coating the Membrane with PAMAM-PEG Network

A rectangular section of approximately 19 cm x 23 cm of a FilmTec RO membrane was cut from continuous roll stock and immersed in a DI water bath for five minutes to remove the protective glycerin coating. It was then placed in a second bath of DI water for subsequent cleaning. Following this, it was affixed to a glass plate of similar dimensions via masking tape at the four corners making sure that it was flat against the plate. DI water was sprayed between the membrane and the glass plate to ensure that the uncoated polyester "bottom" side of the membrane was always in contact with water. The exposed membrane's polyamide "top" surface was also washed with DI water. Coating solutions were prepared with varying weight percents of PAMAM G2.0 and stoichiometric equivalents of PEG-DGE 526 in water. The solution was poured onto the membrane to cover the entire surface and allowed to crosslink for 30 seconds in a stationary horizontal position, and then an aluminum roller was rolled over the membrane twice with little to no vertical pressure to remove excess solution. A pinhole-free coating coverage was determined by using a dye solution and UV detection (*vide infra*).

General procedures for evaluation of coated RO membranes

Flux and Rejection Measurements

Eighty gallons of 1,000 ppm aqueous NaCl (pH ~6.3) in DI water was placed in a reservoir connected to a SEPA CF II (GE Osmonics) test cell system. The membrane was exposed to the solution for 15 minutes at ambient pressure. Following this, a pressure of 100 psi was applied and the first measurement was recorded thirty minutes following the application of pressure. Repeated measurements were taken every thirty minutes after the first reading and over a period of several hours. Flux was reported as the number of milliliters of permeate per minute, rejection measured as total dissolved solids (TDS) in ppm and conductance in micro siemens for the permeate and the reservoir water, and reported as a percent of TDS and conductance level lost in the permeate water.

Hydrophilicity Measurements

The DI water contact angle of the coated membranes was measured using a KSV Instruments CAM 200. Dry membranes were placed on the test stand and affixed via two-sided adhesive tape. A 0.01 mL drop of water was metered out on the dispenser tip and slowly brought down until the drop touched the membrane, whereupon it was pulled away from the dispenser tip via the hydrophilic character of the membrane. Contact angles were measured in one of two ways. In one method, dynamic contact angle measurements were obtained wherein images were taken rapidly, 33 milliseconds apart, and then slowed down to one-second intervals when the drop shape stabilized. In the second method, a single image was captured five seconds after the water droplet was placed in contact with the membrane. Contact angles measured by the second method are being reported in this paper. Contact angles were calculated using the Young-Laplace fitting equation under conditions which allowed for tilt and an automatic baseline.

Results and Discussions

The overall goal of this program was to achieve as high as possible hydrophilization of dynamic RO membrane surfaces, which in turn was expected to result in significantly reduced biofouling under the operative conditions. For this purpose, HB polyamide with a succinate spacer (HB-PA) or PAMAM G2.0 were used as one of the components of network coatings with a PEG derivative. On FilmTec's Low Energy (LE) and Extra Low Energy (XLE) RO membranes as substrates, polymeric coatings with varying relative concentrations of components were prepared. Evaluations of water flux and salt rejection were carried out to determine the effect of surface modifications on the membrane's dynamic properties.

Synthesis of Hyperbranched Polyamides (HB-PA) with NH₂ End-Groups

Using an A_2-type diester, dimethyl succinate, and a B_3-type triamine, tris(2-aminoethyl)amine (TREN), amine-terminated HB-PA was prepared as shown in Reaction Scheme 1. The reaction was carried out with an excess of TREN to obtain the primary amine end-groups. The HB polyamide had a M_w of ~4000 as determined by SEC and TGA analysis showed that it was stable up to about 200°C.

Synthesis of N-hydroxysuccinimide Derivative of MPEG Acetic Acid

The synthesis of the N-hydroxysuccinimide derivative of methoxypolyethyleneglycol acetic acid is illustrated in Reaction Scheme 2. The carboxylic group of MPEG acetic acid was activated by converting it to N-hydroxysuccinimide ester. The reaction was monitored by FTIR (disappearance of –OH band at 3300 cm^{-1}). The yield of the final product was 95%.

Synthesis of MPEG-PAMAM G2.0

The synthesis of MPEG-PAMAM G2.0 is illustrated in Reaction Scheme 3. PAMAM G2.0 contains sixteen exo-presented terminal amine groups. Out of these, eight (50%) were functionalized by reacting with the N-hydroxysuccinimide ester of MPEG. The reaction occurred at room temperature and the reaction was monitored by FTIR until all of the ester groups were consumed. The product was purified by ultrafiltration and the formation of PEG functionalized PAMAM G2.0 product was ascertained by NMR.

Preparation of Dendritic Polymer Network Coatings

Three main types of dendritic polymer network coatings were prepared from HBPs and PAMAM G2.0 as shown in Reaction Scheme 4-6, respectively. The density of crosslinking in PAMAM coatings was varied by utilizing either all (Scheme 4 and 5) or only half (Scheme 6) of their end-groups. The networks were prepared directly on the RO membrane surfaces, where crosslinking started as soon as thin layers of reagent solutions were applied.

Preparation of Polymeric Coatings Using PAMAM G2.0-PEG Brush

In these preparations MPEG-PAMAM G2.0 were crosslinked with PEG-DGE 526 to provide a network coating that contained PEG brushes on its surface. The reaction scheme for the formation of these networks is shown in Scheme 7.

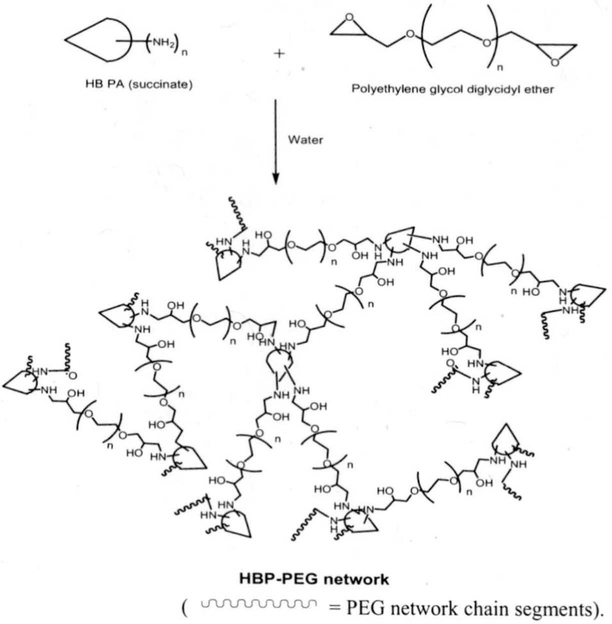

198

HBP-PEG network

(⌇⌇⌇⌇⌇⌇ = PEG network chain segments).

Scheme 4. Preparation of HBP-PEG network with no free terminal amines.

------ = PEG network bonds

PAMAM G2-PEG network

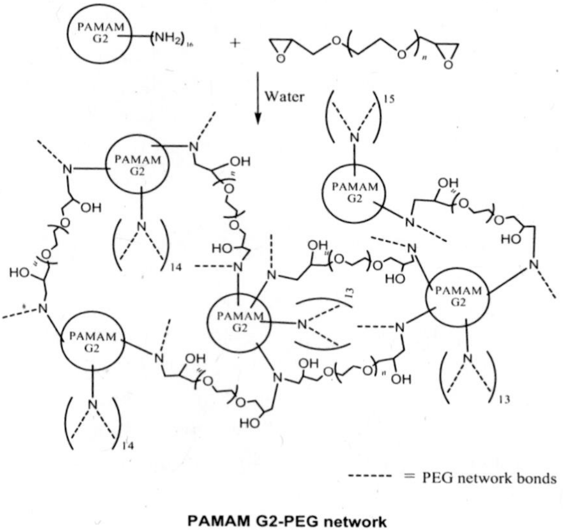

Scheme 5. Preparation of PAMAM G2.0-PEG network with no free terminal amines.

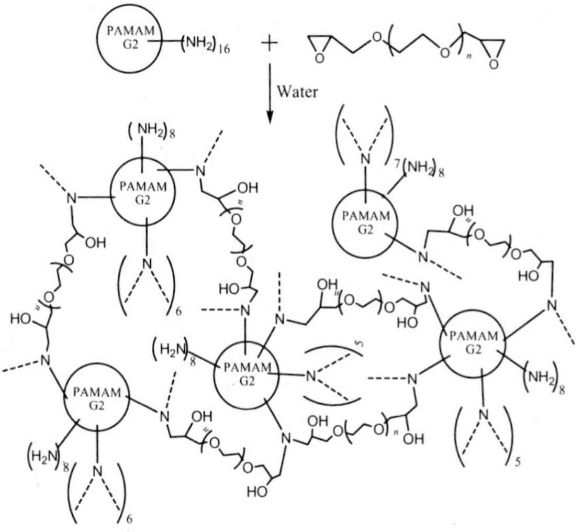

(---- = PEG network chain segments).

Scheme 6. Preparation of PAMAM G2.0-PEG network with 50% free terminal amines.

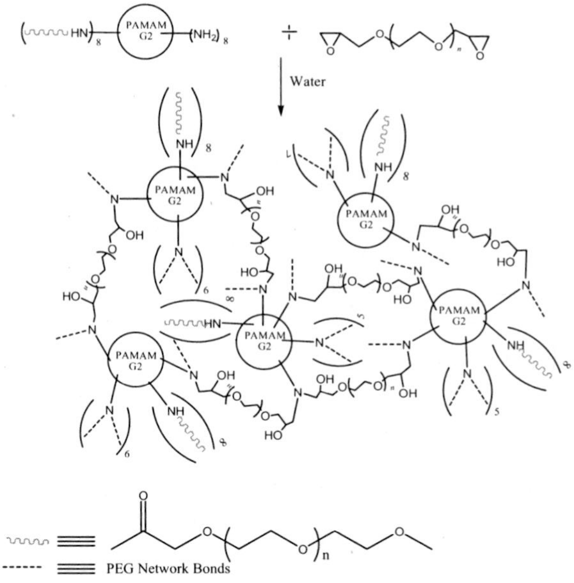

Scheme 7. Preparation of PAMAM G2.0-PEG brush-PEG network.

The idea behind these brush-containing coatings was to prevent biofouling of the membranes by not allowing the biofoulants to come in direct contact with the membrane surfaces by not allowing them to settle there because of the high level of rapid Brownian mobility of the brushes.

Coating coverage evaluations

The benefits of surface modifications are maximized and the costs are minimized when the lowest concentration affording complete coverage is used. A protocol to determine the minimum membrane coverage for varying concentrations consisted of identical procedures for the synthesis of the networks. Analyses of experimental data yielded several relationships between concentration of casting solution and membrane properties: (1) concentration and membrane coverage are directly proportional, (2) concentration and flux are inversely proportional, and (3) the water contact angle is independent of concentration. Following 4 hours of air drying, 254 nm UV light was used to evaluate the surface coverage. Under this light, the coating exhibited a faint green fluorescence while the polyamide surface was blue. It was found that a 0.5 wt. % solution of dendritic polymer yielded a satisfactory coating with uniform coverage.

Surface morphology

A number of samples were evaluated by scanning electron microscopy (SEM) to visualize the eventual changes in surface morphology of membranes after coating. It was found that both the original polyamide surface and the coated surface were quite similar in terms of their physical homogeneity (Figures 5 and 6). In fact, physical irregularities were more pronounced on the polyamide surface than on the polymer-coated surface. Although he improved smoothness of the coated membranes may be expected to negatively contribute to antifouling characteristics, any such change should be dwarfed by the increase in hydrophilicity, which remains the fundamental mechanism to introduce enhanced antifouling capabilities.

Flux and rejection measurements

A SEPA CF II dual cell system set-up was used for flux and rejection measurements. The set-up was equipped with two SEPA CF II cells. The two cells could be used in parallel, allowing comparative evaluations of pairs of membranes. The set-up was further equipped with an automated data collection system for obtaining permeate flux and salt rejection values as a function of time. The dual cell set-up was calibrated by using standard FilmTec LE RO membranes with known performance parameters.

Figure 5. SEM micrograph of the polyamide surface of an uncoated LE membrane.

Figure 6. SEM micrograph of an LE membrane coated with 3 wt.% PAMAM G2.0 and a stoichiometric amount of PEG-DGE in water with no free terminal amines. The surface became slightly more smoother after coating.

Hydrophilicity measurements

Initial assessments of the advancing water contact angles (AWCA) on various dendritic polymers-PEG-526 network coatings indicated that the test protocol required careful optimization. Contact angles measured on membranes dried for 2 hours just before the test were significantly higher than those obtained from membranes that had been dried overnight (>10 hours). The later contact angle measurement process, i.e., using overnight dried membranes, was selected for this study.

The contact angles for the surfaces modified with HBP-PEG coatings were by ~29° lower than those of the uncoated membranes (60°). The decrease in contact angles values for the PAMAM based coatings was even more dramatic (60° to 18°). Compared to the uncoated membrane surfaces, the coated membranes exhibited significantly higher hydrophilicity. The presence or absence of free amines on the surface had little effect on their hydrophilic behavior.

Evaluation of FilmTec RO membranes for selectivity

The salt rejection of the uncoated LE and XLE RO membranes was over 98% at 100 psi operating pressure. The measured fluxes were 13 mL/min and 15 mL/min for the LE and XLE membranes, respectively. These values agreed well with the manufacturer's specifications.

Tables II and III summarize experimental results obtained for the surface modified LE and XLE membranes with HBP-PEG polymer network coatings, respectively. The surface modification chemistry significantly increased the hydrophilic character of the commercial membrane surfaces. This increase was more pronounced for the LE membranes (AWCA = 43°) than for the XLE membrane (AWCA = 31°). The flux and selectivity were not compromised and remained within reasonable efficiency levels as compared to the uncoated membranes.

Tables IV and V summarize experimental results obtained for the surface modified XLE membranes with two different types of PAMAM based polymer network coatings. The surface modification chemistry significantly increased the hydrophilic character of the membrane surfaces by about 25 to 30°. The membrane coating containing the PAMAM G2.0-PEG-DGE (526) network with 50% of the terminal amine groups left unreacted (see Table IV) provided a 25° reduction of contact angle over the uncoated membranes without compromising the salt exclusion. The flux was reduced by about 20% but was well within the benchmark level reported by various research and industrial groups (3).

Table II. Flux, exclusion, and contact angle measurements for FilmTec LE RO membranes surface modified with 1.0 wt.% solution of a HB-PA-PEG coating. The flux and exclusion numbers are the average of two measurements. The contact angle numbers are the average of eight to ten measurements.

Coating	Sample	Permeate Flux (mL/min)[1]	Salt Rejection (%)[1]	Contact angle (°)
Uncoated membrane	--	13.0	98.5	81
1.0% solution of HB-PA-PEG	1	11.2	98.0	39
	2	10.5	97.7	35
	3	11.5	97.8	41
	4	10.1	97.5	37
	Average	**10.8**	**97.8**	**38**

[1]measurements carried out after 6 hours of test

Table III. Flux, exclusion, and contact angle measurements for FilmTec XLE RO membranes surface modified with a 1.0 wt.% solution of a HB-PA-PEG brush-PEG polymer coating. The flux and exclusion numbers are the average of two measurements. The contact angle numbers are the average of eight to ten measurements.

Coating	Sample	Permeate Flux (mL/min)[1]	Salt Rejection (%)[1]	Contact angle (°)
Uncoated membrane	--	14.0	99.0	60
1.0% solution of HB-PA-PEG	1	11.0	98.1	30
	2	10.5	97.7	33
	3	11.1	98.0	31
	4	11.5	96.8	30
	Average	**11.0**	**97.7**	**31**

[1]measurements carried out after 6 hours of test

Table IV. Flux, exclusion, and contact angle measurements for FilmTec
XLE RO membranes surface modified with 0.5 wt.% solution of a PAMAM
G2.0-PEG coating. The flux and exclusion numbers are the average of two
measurements. The contact angle numbers are the average of eight to ten
measurements.

Coating	Sample	Permeate Flux (mL/min)[1]	Salt Rejection (%)[1]	Contact angle (°)
Uncoated membrane	--	14.0	99.0	60
G2.0 PAMAM-PEG-DGE (526) with 50% free amines	1	11.1	99.0	34
	2	12.1	98.0	34
	3	10.6	99.0	33
	4	10.4	99.0	36
	5	11.9	99.0	36
	Average	**11.2**	**98.8**	**35**

[1]measurements carried out after 6 hours of test

Table V. Flux, exclusion, and contact angle measurements for FilmTec XLE
RO membranes surface modified with a 0.5 wt.% solution of a PAMAM
G2.0-PEG brush-PEG polymer coating. The flux and exclusion numbers
are the average of two measurements. The contact angle numbers are the
average of eight to ten measurements.

Coating	Sample	Permeate Flux (mL/min)[1]	Salt Rejection (%)[1]	Contact angle (°)
Uncoated membrane	--	14.0	99.0	60
PAMAM G2.0-PEG brush-PEG DGE (526)	1	13.0	99.0	15
	2	10.0	99.0	19
	3	11.4	97.0	15
	4	10.5	98.0	19
	5	10.0	99.0	18
	6	11.3	99.0	21
	Average	**11.0**	**98.5**	**18**

[1]measurements carried out after 6 hours of test

**Table VI. Effect of hydrophilic dendritic coatings on the separation
properties of commercial RO membranes.**

Sample	Mean contact angle (°)	Salt rejection (%)	Permeate Flux (mL/min)
Uncoated commercial membrane	60 ± 10	99	14.0 ± 2.0
Membrane w/dendritic base coating	35 ± 2	98.8 ± 0.8	11.2 ± 0.9
Membrane w/dynamic (brush) dendritic coating	18 ± 2	98.5 ± 0.8	11.0 ± 0.9

When the entire number of terminal PAMAM dendrimers amine groups was reacted to introduce PEG polymer brushes in the coating network, the results were more dramatic as far as the surface hydrophilicity was concerned (see Table V). The reduction of contact angle was a whopping 42° over the uncoated membranes with negligible reduction in the percentage salt exclusion. The flux was again reduced by 20%.

The overall dynamic properties of LE RO membranes with various coating formulations are summarized in Table VI. Analysis of the data strongly suggests that hydrophilicity can be controlled by manipulating the chemistry of membrane surface coating. PEG brushes play a significant role in enhancing the hydrophilicity and expected to bode well for efficient antifouling without compromising the dynamic properties of the resultant membranes.

Conclusions

New process of surface modifications of aromatic polyamide RO membranes with highly hydrophilic dendritic polymers were developed to enhance antifouling properties without compromising the separation properties of the resulting membranes. The coating thickness was optimized for the best separation properties while providing uniform pinhole free coverage. Evaluations of surface contact angles, permeate fluxes and salt rejections were carried out. A two-pronged approach was implemented. As the first line of defense, a membrane surface with enhanced hydrophilic characteristics prevents most of the biofoulants from settling. The second line of defense consists of hydrophilic brushes on the membrane surfaces that unsettle any biofoulants that get past the first line of defense.

It was found that all applied surface treatments resulted in increased surface hydrophilicity (lower advancing water contact angles) without any detrimental effect on salt rejection and with acceptable permeate flux reduction. The best combination of properties was obtained for 0.5 wt.% solution of a PAMAM G2.0-PEG brush-PEG polymer coating which led to 42° AWCA and only 20% permeate flux reduction.

Acknowledgements

The authors gratefully acknowledge funding from the Department of Defense under Contract Numbers W9115R-05-C-0026 and W9115R-07-C-0036. The RO membranes were graciously provided by FilmTec Corporation.

References

1. Murakami, S.; Senoo K.; Toki, S. *Polymer* **2002**, *43*, 2117.
2. Toki, S.; Sics, I.; Ran, S.; Liu, L.; Hsiao, B.S.; Murakami, S. *Macromolecules* **2002**, *35*, 6578.
3. Toki, S.; Sics, I.; Ran, S.; Liu, L.; Hsiao, B.S. *Polymer* **2003**, *44*, 6003.
4. Toki, S.; Hsiao, B.S. *Macromolecules* **2003**, *36*, 5915.
5. Trabelsi, S.; Alobouy, P.A.; Rault, J. *Macromolecules* **2003**, *36*, 7624.
6. Kinghorn, P.H.; Haas, W.E. *International Water Conference, 61st Annual Meeting*, Oct. 22-26, 2002, p. 276.
7. *Reverse Osmosis: Membrane Technology, Water Chemistry, and Industrial Applications;* Amjad, Z., ed.; Van Nostrand Reinhold: New York, 1993.
8. *Perry's Chemical Engineers' Handbook;* Perry, R.H.; Green, D.W., Eds.; McGraw-Hill: New York, 7th ed., 1997.
9. Baker, R.W. *Membrane Technology and Applications;* John Wiley & Sons Ltd.: Chichester, 2nd Ed., 2004.
10. Pinnau, I.; Freeman, B.D. in *Formation and Modification of Polymeric Membranes;* Pinnau, I.; Freeman, B.D., Eds.; ACS Symposium Series 744; American Chemical Society: Washington DC, 2000; p 1.
11. Riley, R.L. in *Membrane Separation Systems – A Research & Development Needs Assessment;* Baker, R.W.; Cussler, E.L.; Eykamp, W.; Koros, W.J.; Riley, R.L.; Strathmann, H., Eds.; Department of Energy, Publication number DOE/ER/30133-H1: Springfield, VA, 1990, p. 5.
12. Koo, Ja-Young; Hong, S.P.; Kang, J.W.; Kim, N., *US Patent Appln. No. 0,121,844 A1*, **2003**.
13. Hachisuka, H.; Ikeda, K., *US Patent Appln. No. 6,177,011 B1*, **2001**.
14. http://www.dow.com/liquidseps/prod/app_lelp.htm
15. Sarkar, A.; Carver, P.I.; Zhang, T.; Merrington, A.; Bruza, K.J.; Rousseau, J.L.; Keinath, S.E.; Dvornic, P.R. *J. Membr. Sci.* **2009**, *submitted*.
16. Sarkar, A.; Carver, P.I.; Zhang, T.; Merrington, A.; Bruza, K.J.; Rousseau, J.L.; Keinath, S.E.; Dvornic, P.R. *Desalination* **2009**, *submitted*.
17. Newkome, G.R.; He, E.; Moorfield, C.N., *Chem. Rev.* **1999**, *99*, 1689.
18. Dvornic, P.R.; Tomalia, D.A. in *"The Polymeric Materials Encyclopedia,"* vol. 3, Salamone, J.C. (Ed.), CRC Press. Inc., Boca Raton (FL), 1996, p.1814.
19. Dvornic, P.R.; Tomalia, D.A., *Curr. Opin. Coll. Interface Sci.* **1996**, *1*, 221.
20. Dvornic, P.R.; Hu. J.; Meier, D.J.; Nowak, R.M., "Hyperbranched polycarbosilanes, polycarbosiloxanes, polycarbosilazanes and copolymers thereof," *US Patent No. 6,384,172*, **2002**.

21. Dvornic, P.R.; Hu, J.; Meier, D.J.; Nowak, R.M., "Hyperbranched polyureas, polyurethanes, polyamidoamines, polyamides and polyesters," *US Patent No. 6,534,600 B2*, **2003**.

22. Dvornic, P.R.; Hu, J.; Meier, D.J.; Nowak, R.M., "Hyperbranched polymers with latent functionality and methods of making same," *US Patent No.* 6,646,089 B2, **2003**.

23. Dvornic, P.R.; Hu, J.; Meier, D.J.; Nowak, R.M., "Silicon-Containing Dendritic Polymers by bimolecular polymerization," *Polym. Prepr.,* **2004**, *45(1)*, 585.

24. Dvornic, P.R. and Meier, D.J., "Silicon-Containing Dendritic Polymers by Bimolecular Nonlinear Polymerization," in *Silicon-Containing Dendritic Polymers*; Dvornic, P.R.; Owen, M.J., Eds.; Springer: Gylford, UK, 2009, p.401.

25. Wang, J.; Cheng, Y.; Xu, T., *Recent Patents on Chem. Engn.*, **2008**, *1*, 41.

Chapter 13

High Throughput Screening to Modify Surface Properties and Obtain High Performance Membranes

Mingyan Zhou[1], James E. Kilduff[1], and Georges Belfort[2]

[1]Department of Civil and Environmental Engineering
[2]Department of Chemical and Biological Engineering
Rensselaer Polytechnic Institute, Troy, NY 12180

A novel high throughput method for synthesis and screening of feed-specific fouling-resistant surfaces was developed. The method combines a high throughput platform (HTP) together with photo (UV)-assisted graft polymerization (PGP) of vinyl monomers to commercial poly(ether sulfone) (PES) membranes. This new HTP-PGP method was used to discover new surfaces able to resist membrane fouling by natural organic matter (NOM) and bovine serum albumin (BSA). Several surfaces, including grafted amides, amines, and poly(ethylene glycol) methyl ether methacrylates (PEG-MAs) produced excellent surfaces for both feeds. Grafted zwitterion surfaces appeared to work better for NOM feeds. With a few exceptions, our findings are consistent with known attributes of protein-resistant surfaces. Such exceptions include grafted quaternary amine and grafted carboxylic monomer surfaces that worked well for NOM (but not for BSA) and an aromatic monomer worked well for BSA but not for NOM.

Introduction

Customizing membrane surfaces for particular feeds is urgently needed to minimize the interaction of feed components (dissolved solutes and suspended particles) with membrane surface with the goal of minimizing flux decline and maximizing flux recovery after cleaning. UV-assisted graft polymerization of vinyl monomers to commercial poly(aryl sulfone) membranes offers an attractive route to customizing membrane surfaces. To date, such surface modification has been implemented at the bench scale (1, 2), producing and testing new surfaces in a linear and sequential manner. Therefore, the process has been slow, with a low probability of success and with little mechanistic insight. As a result, developing new polymeric materials with appropriate surface or functional characteristics for different filtration applications has involved great effort and expense, and has taken many years.

High throughput techniques are powerful tools to quickly screen a variety of variables, such as compounds, reactions, operating conditions, and parameters. Such techniques have been widely and successfully used in chemistry and biology (3-5). In recent years, researchers have started to adapt this method to the membrane field. However, until now applications have been limited to optimizing process variables in the development of filtration processes using commercial available filtration membranes (6, 7), and optimizing membrane casting dope composition for some specific feeds (8-10).

Here, we adapt a high throughput platform (HTP) approach to the facile modification of poly(aryl sulfone) membranes, using a HTP together with our patented photo-induced graft polymerization (PGP) method. In the PGP method, depicted schematically in Figure 1, poly(ether sulfone) (PES) membranes are UV-irradiated, cleaving trunk polymer chains and forming reactive radical sites. Either water or ethanol-soluble vinyl monomers covalently bond to these radical sites and undergo free-radical polymerization (2). We call the combined method HTP-PGP. The novel method proposed here is inexpensive, fast, simple, reproducible and scalable way to synthesize and screen fouling-resistance surfaces by modifying poly(aryl sulfone), which has excellent physical and transport characteristics but poor surface chemistry.

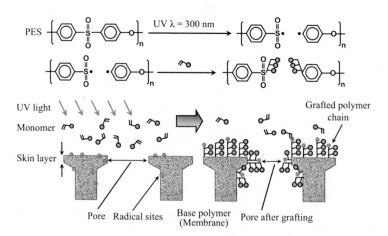

Figure 1. In the PGP method, poly(aryl sulfone) membranes are UV-irradiated (λ ≈ 300 nm), cleaving trunk polymer chains and forming reactive radical sites. Vinyl monomers chemically bond to these radical sites and undergo free-radical polymerization. After Yamagishi et al. (2).

Our approach involved selecting 66 commercially available monomers to create an initial monomer library of likely candidates. These monomers were then employed to modify PES surfaces using the HTP-PGP approach. Candidate surfaces were synthesized, characterized in terms of permeability, challenged separately by static adsorption of NOM and BSA solutions, and screened by subsequent water filtration in the same multi-well filter plate. HTP was used to discover new surfaces for applications involving NOM and BSA filtration. Grafted surfaces were challenged with a static adsorption assay, chosen because in preliminary experiments with BSA feeds it correlated with results from a filtration assay.

Materials and Methods

Materials

Membrane

Polypropylene 96-well filter plates (Seahorse Labware, Chicopee, MA) were used in HTP-PGP experiments. A 100 kDa cut-off PES membrane coupon (effective area 19.35 mm^2) was mounted by the manufacturer on the bottom of each 400-μL well. The hydraulic resistance of the 96 membranes ranged from 8.12×10^{11} to 9.49×10^{11} m^{-1} with a coefficient of variation equal to 4.0%.

Monomers

Commercial vinyl monomers (66 total) were purchased from Sigma-Aldrich (Saint Louis, MO) and were used as-received without further purification. These monomers were classified into 9 groups: hydrophobic methacrylates (HPO MAs, methacrylates having alkane chains or rings), hetero ring group monomers (having an oxygen or nitrogen-containing ring), aromatic monomers, hydroxy monomers (containing multiple OH groups), poly(ethylene glycol) (PEG) monomers (containing ethylene glycol repeating unit), strong and weak acid monomers (containing either carboxylic or sulfonic groups), amine monomers, basic and zwitterionic (zwit) monomers (basic monomers and zwitterions), and some other monomers which could not be classified into the above groups. A monomer concentration of 0.2 M was employed for grafting experiments. These monomers were either dissolved in reagent grade water or ethanol depending on their solubility.

Feed Solutions

Elliott Humic Acid (EHA) from International Humic Substances Society (St. Paul, MN) was applied as a model NOM. A solution containing 50 mg/L EHA (as TOC) and 0.01 M ionic strength at pH 7 was used in HTP-PGP static adsorption experiments. The ionic strength and pH were adjusted by adding NaCl solid, and HCl and NaOH solutions respectively.

Bovine serum albumin was chosen as a model protein to assess membrane fouling. BSA (MW = 67 kDa, pI = 4.7) is negatively charged under our experimental conditions. Solution was prepared by dissolving BSA into phosphate buffered saline (PBS) solution to yield a protein concentration of 1 g/L. PBS buffer solution contained 10 mM phosphate buffer, 2.7 mM potassium chloride and 137 mM sodium chloride with pH 7.4 at 25 °C. BSA and PBS tablets were purchased from Sigma-Aldrich (Saint Louis, MO).

Methods

Preparation of Modified Surfaces

The membranes on the 96-well filter plates were modified using the UV-induced graft polymerization method. The approach is shown schematically in Figure 2. UV irradiation was conducted in a chamber (F300S, Fushion UV Systems, Inc. Gaithersburg, MD) containing an electrodeless microwave lamp (~ 7% of the energy was at < 280 nm). A bandpass UV filter (UG-11, Newport Corporation, Franklin, MA) was placed between the 96-well filter plate and the UV lamp to reduce the energy at wavelengths below 280 nm to < 1%.

The membrane modification consisted of the following steps. After washing, the hydraulic permeability of each well was measured simultaneously with DI water. The membranes were then modified by adding monomer solution (200 µL) to each well, shaking the plates on an orbital shaker at 100 rpm for 1 hr, reducing O_2 level by purging with N_2 for 15 min, and irradiating plates in the UV chamber for 30 s. After modification, the plates were washed by shaking in DI for 1 hr. Each monomer was evaluated with four replicates. Four membrane coupons were treated with ethanol without UV irradiation to serve as a control for membranes grafted with the monomers dissolved in ethanol, and another four membranes were used as-received to serve as a control for the membranes grafted with monomers dissolved in water.

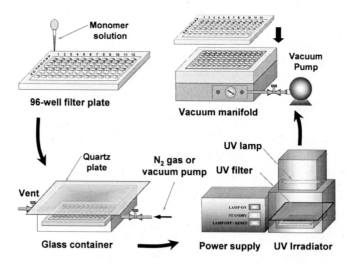

Figure 2. Schematic illustration of HTP-PGP experimental setup and approach. After Zhou et al. (11).

Evaluation of Modified Membranes by Static Adsorption

The fouling potential of modified and control membranes was evaluated using a static adsorption protocol. The water permeability after adsorption was measured as a criterion for membrane performance. In this method, 300 μL of foulant solution was added to each well, and the plate was sealed with adhesive film to eliminate evaporation. The plate was then placed on a shaker (as above) for 44 hrs. After equilibration, the wells were then gently emptied, and DI flux was measured. The membrane resistance was calculated using flux values. The resistance increase of the modified membranes caused by foulant adsorption was compared with that of control membranes to evaluate foulant/surface interactions.

Analytical Methods

A Microplate Spectrophotometer (PowerWave XS, BioTek Instruments Inc., Winooski, VT) was used to measure the volume of permeate solution in the receiver plate wells. The acrylic 96-well receiver plates allow permeate analysis by light absorbance in near infrared region. The volume of permeate in each receiver well was measured at 977 nm. NOM and BSA do not absorb at this wavelength, whereas water exhibits an absorbance peak. Volumetric flux, J_v (m/s) was calculated as $J_v = V/At$, where V (m³) is the cumulative permeate volume, A (m²) is the membrane area, and t (s) is the filtration time. The resistance of membrane was calculated from $R = \Delta P/\mu J_v$, where ΔP (Pa) is the transmembrane pressure, μ (g/m s) is the solution viscosity at $22 \pm 1°C$.

Results and Discussion

To assess foulant/surface interactions, a fouling index, $\mathfrak{R}$, was calculated as the resistance increase of grafted membranes caused by fouling normalized by that of ungrafted membrane control, $\mathfrak{R} = \Delta R_{mod}/\Delta R_{control}$, where $\Delta R_{mod} = (R_{fouled} - R)_{mod}$ and $\Delta R_{control} = (R_{fouled} - R)_{control}$. The control was the as-received membrane treated with either water or ethanol, depending on which was used to dissolve the monomer. The increase in the modified membrane resistance after foulant adsorption should be lower than that of the control when the modified surface resists foulant interactions. In other words, a fouling index lower than 1 indicates the produced new surfaces are better than PES materials in terms of resisting static adsorption of the feed components.

The fouling index of the best fouling-resistant surfaces is shown in Figure 3 for both NOM and BSA feeds. Of the top performers, 8 worked well for both feeds, including 4 amines (#s 53, 51, 55, 52), 2 PEGs (#34 and #35), an HPO MA (#7), and a zwitterionic monomer (#59). Several monomers performed well for NOM but did were not among the top performers for BSA. These include a zwitterion (#60), a basic monomer (#61), and a carboxylic acid monomer (#45).

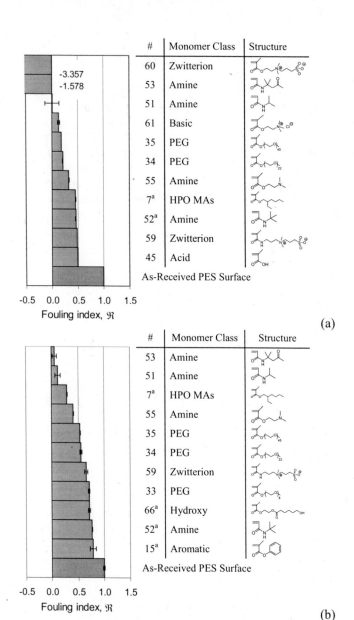

Figure 3. Top fouling-resistant surfaces for NOM (a) and BSA (b) from a total of 66 commercial monomers relative to the as-received PES membrane using the HTP-PGP method. Success is measured in terms of fouling index, $\mathfrak{R}$. Superscript "a" indentifies ethanol soluble monomers. #s represent the monomer number in the library of 66 monomers.

In addition, several monomers performed well for BSA, but did not produce surfaces that were among the best performers at resisting NOM fouling. These included a short-chain PEG (#33) a hydroxy monomer (#66) and an aromatic monomer (#15).

With some exceptions, our findings are generally consistent with results from studies of protein interactions with surfaces having a variety of functionality created using self-assembled monolayers (SAMs) of alkanethiolates on gold as a model substrate (12-19). Such studies have identified general features of surfaces having low affinity for proteins: (i) they are hydrophilic (wettable), (ii) they contain hydrogen bond acceptors, (iii) they lack hydrogen bond donors, (iv) they are electrically neutral (12-15).

Among the surfaces that mitigated fouling for both NOM and BSA feeds, the long-chain PEG monomers (#34 and #35), represent the "standard" for protein resistance, and satisfy all four of the above criteria. However, the best performing monomer for BSA, and the second best for NOM, diacetone acrylamide (#53), contains a secondary amine which can act as a hydrogen bond donor. It is likely that the location of the amine group adjacent to the carbonyl oxygen limits its reactivity. Others have noted that primary and secondary amines adsorb more protein than structurally similar groups in the form of amides (15). Furthermore, it should also be noted that other molecules containing hydrogen bond donors, such as mannitol, have exhibited protein resistance (20). Three other monomers containing amine groups also performed well. The 2-(dimethylamino) ethyl methacrylate (#55) contains a tertiary amine, the N-isopropylacrylamide (#51) and the N-tert-butylacrylamide (#52) also contains amide groups. The zwitterion [3-(methacryloylamino)propyl] dimethyl(3-sulfopropyl)ammonium hydroxide inner salt (#59) conforms to the net neutrality criterion, but also contains a secondary amine in an amide group. The most surprising monomer was the 2-ethylhexyl methacrylate (#7), which is much less hydrophilic than the other high performers.

For the BSA feed-specific fouling resistant surfaces, the PEG (#33) also satisfies all the four criteria. However, this PEG has only 8 ethylene glycol repeating units as compared with 22 for #34 and 45 for #35; therefore it is likely that the produced graft chains were shorter after surface modification, which may explain why monomer #33 did not work well for NOM. The caprolactone 2-(methacryloyloxy)ethyl ester monomer (#66) terminates in an alcohol group, separated from the hydrophilic portion of the molecule by a five-carbon chain. It is possible that this chain length is long enough to promote self-association and reduce the hydrogen bond donor reactivity with proteins. The success of aromatic monomer #15, phenyl methacrylate, was unexpected, because it is not as hydrophilic as other top performers.

For the NOM feed-specific fouling resistant surfaces, the zwitterion [2-(methacryloyloxy)ethyl]dimethyl-(3-sulfopropyl)ammonium hydroxide (#60) conforms to the net neutrality criterion. It performed the best for NOM; however, it was not ranked in top performers for BSA feed. Interestingly, a good performer for NOM feed was the basic monomer [2-(methacryloyloxy)ethyl] trimethylammonium chloride (#61). It was positively charged under our experimental conditions; therefore, it does not satisfy the general features of surfaces having low affinity for proteins. However, this monomer ranked the fourth for NOM fouling resistant surfaces. The methacrylic (carboxylic) acid monomer (#45) performed well for NOM, likely because it induced charge repulsion between the grafted surface and the like-charged feed components. However, this monomer did not perform well for BSA, which was expected based on the general criteria for protein resistance.

Conclusions

A novel high throughput method for synthesis and screening of customized fouling-resistant surfaces was developed by combining a high throughput platform approach together with our patented photo-induced graft polymerization method, to allow facile modification of commercial poly(aryl sulfone) membranes. This method is inexpensive, fast and simple at discovering feed-specific fouling-resistant surfaces by static adsorption.

The HTP approach was employed in a discovery mode to identify many surfaces from a library of 66 monomers that perform significantly better than the as-received membrane, offering significantly lower resistance due to fouling of NOM and BSA. Our results are generally consistent with rules governing general features of surfaces having low affinity for proteins. The grafted amines (especially in the amide forms), and long chain PEGs produced excellent surfaces for both feeds. Grafted zwitterion surfaces appeared to work better for NOM feed. Some exception was also found for NOM feed: the grafted basic and acid surfaces worked well for NOM but not for BSA.

Future work will involve compiling foulant/surface interaction data, membrane performance parameters, and monomer structural features as input to develop structure/property relationships (QSPR), which will provide mechanistic insight that can be used for designing new surfaces for particular feed solutions, and for expanding the initial monomer library.

Acknowledgements

The authors acknowledge the U.S. Environmental Protection Agency (EPA grant RD83090901-0), U.S. National Science Foundation (CBET-0730449) for financial support, Seahorse Labware (Chicopee, MA) for filter plates, and Millipore Corporation (Bedford, MA) for membranes.

References

1. Crivello, J. V.; Belfort, G.; Yamagishi, H. In *US Patent Number 5,468,390*; Rensselaer Polytechnic Institute, Troy, N.Y.: United States, **1995**.
2. Yamagishi, H.; Crivello, J. V.; Belfort, G. Development of a novel photochemical technique for modifying poly (arylsulfone) ultrafiltration membranes. *Journal of Membrane Science* **1995**, *105*, 237-247.
3. Bannwarth, W.; Felder, E. *Combinatorial Chemistry: A Practical Approach* Wiley-VCH Weinheim, Germany, **2000**.
4. Gold, L.; Brown, D.; He, Y.-Y.; Shtatland, T.; Singer, B. S.; Wu, Y. From oligonucleotide shapes to genomic SELEX: Novel biological regulatory loops. *Proceedings of the National Academy of Sciences of the United States of America* **1997**, *94*, 59.
5. Clackson, T.; Hoogenboom, H. R.; Griffiths, A. D.; Winter, G. Making antibody fragments using phage display libraries. *Nature* **1991**, *352*, 624.
6. Jackson, N. B.; Liddell, J. M.; Lye, G. J. An automated microscale technique for the quantitative and parallel analysis of microfiltration operations. *Journal of Membrane Science* **2006**, *276*, 31-41.
7. Chandler, M.; Zydney, A. High throughput screening for membrane process development. *Journal of Membrane Science* **2004**, *237*, 181-188.
8. Vandezande, P.; Gevers, L. E. M.; Paul, J. S.; Vankelecom, I. F. J.; Jacobs, P. A. High throughput screening for rapid development of membranes and membrane processes. *Journal of Membrane Science* **2005**, *250*, 305-310.
9. Bulut, M.; Gevers, L. E. M.; Paul, J. S.; Vankelecom, I. F. J.; Jacobs, P. A. Directed Development of High-Performance Membranes via High-Throughput and Combinatorial Strategies. *Journal of Combinatorial Chemistry* **2006**, *8*, 168 - 173.
10. Vandezande, P.; Gevers, L. E. M.; Vankelecom, I. F. J.; Jacobs, P. A. High throughput membrane testing and combinatorial techniques: powerful new instruments for membrane optimisation *Desalination* **2006**, *199*, 395-397.
11. Zhou, M.; Liu, H.; Venkiteshwaran, A.; Kilduff, J.; Anderson, D. G.; Langer, R.; Belfort, G. High Throughput Discovery of New Fouling-Resistant Surfaces. *Submitted* **2008**.
12. Ostuni, E.; Chapman, R. G.; Holmlin, R. E.; Takayama, S.; Whitesides, G. M. A survey of structure-property relationships of surfaces that resist the adsorption of protein. *Langmuir* **2001**, *17*, 5605-5620.
13. Holmlin, R. E.; Chen, X.; Chapman, R. G.; Takayama, S.; Whitesides, G. M. Zwitterionic SAMs that resist nonspecific adsorption of protein from aqueous buffer. *Langmuir* **2001**, *17*, 2841-2850.
14. Chapman, R. G.; Ostuni, E.; Takayama, S.; Holmlin, R. E.; Yan, L.; Whitesides, G. M. Surveying for surfaces that resist the adsorption of proteins. *Journal of the American Chemical Society* **2000**, *122*, 8303.
15. Chapman, R. G.; Ostuni, E.; Liang, M. N.; Meluleni, G.; Kim, E.; Yan, L.; Pier, G.; Warren, H. S.; Whitesides, G. W. Polymeric thin films that resist the adsorption of proteins and the adhesion of bacteria. *Langmuir* **2001**, *17*, 1225-1233.

16. Ostuni, E.; Yan, L.; Whitesides, G. M. The interaction of proteins and cells with self-assembled monolayers of alkanethiolates on gold and silver *Colloids and Surfaces B: Biointerfaces* **1999**, *15*, 3-30.

17. Vutukuru, S.; Bethi, S. R.; Kane, R. S. Protein interactions with self-assembled monolayers presenting multimodal ligands: A surface plasmon resonance study. *Langmuir* **2006**, *22*, 10152-10156.

18. Chen, S.; Zheng, J.; Li, L.; Jiang, S. Strong resistance of phosphorylcholine self-assembled monolayers to protein adsorption: Insights into nonfouling properties of zwitterionic materials. *Journal of the American Chemical Society* **2005**, *127*, 14473-14478.

19. Mrksich, M.; Whitesides, G. M. Using Self-Assembled Monolayers That Present Oligo(ethylene glycol) Groups to Control the Interactions of Proteins with Surfaces. *ACS Symposium Series* **1997**, *680*, 361-373.

20. Luk, Y.-Y.; Kato, M.; Mrksich, M. Self-assembled monolayers of alkanethiolates presenting mannitol groups are inert to protein adsorption and cell attachment. *Langmuir* **2000**, *16*, 9604-9608.

Chapter 14

TiO$_2$ Nanowire Free-Standing Membrane for Concurrent Filtration and Photocatalytic Oxidation Water Treatment

Xiwang Zhang, Alan Jianhong Du, Jiahong Pan, Yinjie Wang and Darren Delai Sun[1]

School of Civil and Environmental Engineering
Nanyang Technological University, Singapore 639798
[1] Corresponding Author : ddsun@ntu.edu.sg, Tel: +65 6790 6273

A flexible, uniform, and multifunctional TiO$_2$ nanowire membrane was successfully fabricated using a hydrothermal method. XRD results indicate that the TiO$_2$ nanowires forming the membrane consist of anatase TiO$_2$ and Na$_2$(Ti$_6$O$_{13}$) titanate. These nanowires were 20-100 nm in diameter and had a typical length of several micrometers to tens of micrometers. Under UV irradiation, the membrane showed an excellent performance on concurrent filtration and photocatalytic oxidation degradation of HA in water. A removal rate of 100% and 93.6% for Humic acid and TOC were achieved up to, respectively. The nanowire membrane proves to have remarkable potential for water and wastewater purification, and could be used to eliminate other foulants such as microorganisms and trace organics. It could produce a new flexible dye sensitized solar cell (DSSC) as well.

Introduction

The demand for drinking water as a precious resource has increased tremendously in terms of quantity and quality over the past few decades[1]. One of the main strategic solutions recommended for the fulfillment of increasing water demand in quantity and quality is through water reclamation technologies for the conservation and recovery of drinking water. However, the presence of contaminants such as natural organic matters (NOMs), trace organics and microorganisms that accumulate in different sources of raw water creates major problems which make raw water unsuitable for direct human consumption. There is a need for cost effective water treatment prior too water being consumed

The alum coagulation/flocculation together with chlorination and ozone oxidation is a conventional technology that has been widely used in water treatment plants for the removal of particles as well as the large molecular weight contaminants found within raw water. Unfortunately, this technology is unable to completely remove the small molecule weight contaminants, thus generating extra volumes of sludge, which require further treatment and safe disposal. The large infrastructure development requires large land space. Furthermore, aluminum presence in the treated water has been suspected to develop the onset of Alzheimer's diseases[2]. Small molecule weight NOMs such as fulvic acids and humic acids (HA) has been shown to react with the major disinfectants such as chlorine, ozone, chlorine dioxide, chloramines to produce disinfection by-products, such as trihalomethanes (THMs), haloacetic acids (HAAs), bromoform ($CHBr_3$), dibromoacetic acid (DBAA), and 2,4-dibromophenol (2,4-DBP) etc, which are carcinogenic compounds[3].

Conventional micro/ultra filtration (MF/UF) membranes are an advanced water treatment processes for producing high quality drinking water with a small footprint size[4]. However, the commercial filtration membranes that available in the industry have high fouling tendencies caused by the deposition and absorption of the contaminants such as NOMs, trace organics and microorganisms, which are the major problems in using the filtration membranes for producing high quality drinking water. There is urgent need in searching for a new generation of membrane which will be able to overcome the existing problems of the conventional filtration membrane fouling and leaking of contaminants within the membrane pore.

Nanotechnology has great potential for molecular separation applications by offering more desired structurally control materials for such needs. Titanium dioxide (TiO_2) nanosized particle is a popular photocatalyst which attracts much attention from both fundamental research and practical applications for the removal of contaminants from water [5, 6, 7, 8, 9]. It is well known that TiO_2 which functions as a photocatalyst would generate electrons and holes when UV light strikes its crystal surface. While the holes are strong oxidants which than can be applied to various environmental problems, such as removal of the contaminants from water through photocatalytic oxidation reactions [10, 11, 12]. Unfortunately, commercially available TiO_2 photocatalyst which is in the nano size range, has an inherent and significant drawback [13, 14], that is, the difficulty in separation and recovery of the nanosized particle from water. In order to solve the problem of separation and recovery, it is important to re-design the nanostructured TiO_2 photocatalytic material thus not only solving the problem but also further increases the photocatalytic activity of TiO_2

In the past 10 years, our research group has put in a large amount of effort to develop a robust, flexible and free-standing multi-functional TiO_2

membrane using nanofabrication technology to overcome the problems that arose from conventional membranes and nanosized TiO_2 photocatalyst. The membrane can be formed in four different nanostructes which are nanowire, nanofiber, nanorods and nanotube. It was found that a nanowire membrane is the most inexpensive option out of the four nanostructures.

The pioneer work in developing a multi-functional free-standing membrane to overcome the conventional membrane problems was carried out by Anderson and co-workers [15, 16]. They fabricated non-flexible TiO_2 photocatalytic ceramic membrane using sintering of TiO_2 particles on Υ-Al_2O_3 support. Following that many research studies have been taken out to fabricate TiO_2 membranes by coating TiO_2 films on various supports [17-29]. Unfortunately, photocatalytic efficiency of these coated TiO_2 membranes on various supports was lower than that of nanosized TiO_2 photocatalysts. Reduction of the filtration production rate, which permeates through such coating or sintering is another drawback.

Recently nanotechnology research has shown that one dimensional (1D) nanostructured TiO_2, in the form of nanowires, nanofibers, nanorods and nanotubes, can be synthesized by chemical methods[30-36]. Among these chemical methods, hydrothermal method is commonly used due to its un-complex operation. For example P25 TiO_2 powders can be converted into TiO_2 nanowires using a stainless steel pressure vessel (autoclave) under controlled conditions such as temperature /pressure with concentrated NaOH solution. Under these controlled conditions, the Ti-O-Ti bonds were broken and then Ti-O-Na and Ti-OH bonds were formed, which resulted in the formation of TiO_2 nanowires [34,37]. Research results also indicate that TiO_2 nanowires exhibit superior photocatalytic oxidation efficiency compared with P25 TiO_2 due to its larger surface area and presence of quantum size effect (crystal size less than ten nanometer). In the past five years, substantial amount of efforts [38-42] has been devoted to the study of organizing nanowire/fiber/tube into 2D nanomaterials such as carbon nanofilm and strong, transparent, multifunctional carbon nanotube sheet. These 2D nanomaterials brought in new properties such as flexibility, concurrent filtration and multifunctionality while retaining the properties of its 1D nanomaterial.

In this section, we introduce a 2D TiO_2 nanowire membrane as shown in Figure 1 which is fabricated from TiO_2 nanowires suspension synthesized from hydrothermal reaction. The advantages of the TiO_2 nanowire membrane are: (1) full surface exposure to UV or solar for self-regeneration through photocatalytic oxidation reaction, which effectively eliminates the membrane fouling problem; (2) concurrent membrane filtration for separation purpose; (3) high surface area, which allows higher adsorption rate of various trace organics and microorganisms for improving water quality; (4) and higher acid/basic and temperature controlled resistance; (5) environmentally friendly and longer membrane life span; (6) flexible property which enables the membrane to be formed into various membrane modules for larger commercial application. We will also investigate a 2D TiO_2 nanowire membrane by evaluating its permeability and photocatalytic activity.

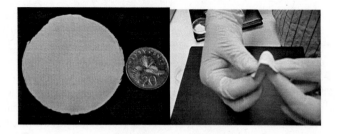

Figure 1. Images of multifunctional, free-standing, flexible TiO$_2$ nanowire membrane.

Experimental Procedure

Synthesis and Characterization of TiO$_2$ nanowire membrane

Synthesis:

0.20 g of TiO$_2$ powder (P25, Degussa) was added in with 30mL of 10M NaOH solution in a 45 mL Teflon-lined container in stainless steel vessel and then it was left in 180°C for hydrothermal reaction for 2 days. At the end of the 2 days reaction, white pulp suspension consisting of long nanowires was produced and collected. Following the collected nanowires were washed with distilled water and diluted hydrochloric acid solution (pH 2) for 3 times. 0.1 (wt) % surfactant (F-127) was then added into the washed white pulp nanowire suspension. To make a nanaowire membrane, a suction-filter fitted with a glass filter (Advantec, GC-50, pore size: 0.45 µm) was used to retain the washed white pulp TiO$_2$ nanowire suspension on its surface, to form a TiO$_2$ nanowire membrane. Residual surfactant left in the TiO$_2$ nanowire membrane was washed away with distilled water using the same suction-filter. After the residue was removed, the TiO$_2$ nanowire membrane was left to dry at room temperature. The glass filter was then removed to form a free-standing TiO$_2$ nanowire membrane prior to calcination in a furnace at 700 °C for 2 h with a ramp of 2 °C/min for crystal phase changes.

Characterization:

The physical properties such as morphologies and the crystal phases of TiO$_2$ nanowire membrane was examined using scanning electron microscopy (SEM), Leica LT7480 and field emission scanning electron microscopy (FESEM, JEOL 6340), powder X-ray diffraction (XRD), Bruker AXS D8 advance and JEOL 2010 transmission electron microscopy (TEM).

Permeability of TiO₂ nanowire membrane

The permeability of the TiO₂ nanowire membrane was evaluated using standard polystyrene microspheres[44, 45], size ranges from 0.05, 0.1, 0.2, 0.5, 1 to 2 μm in diameter obtained from Alfa Aesar. The microspheres was mixed in pure water to obtain 0.0033 (wt) % solution for each Polystyrene microsphere diameter size was filtered in Millipore UF Stirred Cells. TOC analyzer was used to detect the presence of polystyrene microspheres in feed and permeate. The separation factor of the TiO₂ nanowire membrane was than determined using the equation.

$$\text{Seperation Factor} = \left(1 - \frac{C_{permeate}}{C_{feed}}\right) \times 100\%$$

Where $C_{permeate}$ and C_{feed} are the PS solution concentration of permeate collected and the original feed, respectively.

Photocatalytic oxidation and membrane filtration

The concurrent photocatalytic oxidation activity and filtration of TiO₂ nanowire membrane were investigated by using a dead-end filtration equipment together with UV lamp (11 W Upland 3SC9 Pen-ray lamp, 254nm) and auto data collection system in a concurrent continuous operation mode as shown in Figure 2. The volume of the filter cup was 250 mL. The UV light was above the TiO₂ nanowire membrane at a distance of 1cm. HA concentration was measured by determining its absorbance at 436 nm using a UV-visible spectrophotometer. Total organic carbon (TOC) concentration was determined by a Shimadzu TOC-Vcsh TOC analyzer.

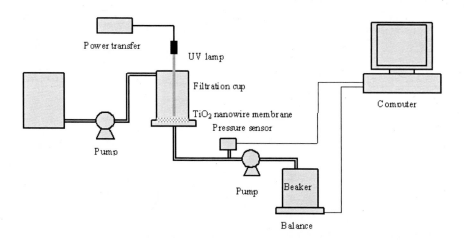

*Figure 2. The set-up for concurrent filtration and photocatalytic oxidation. (Reproduced from reference, Zhang et al, Journal of Membrane Science **2008,** 313, 44-51)*

The photocatalytic oxidation activity of the TiO_2 nanowire membrane was evaluated in the experiment setup in batch operation mode using humic Acid (HA) obtained from Fluka as a model contaminant. 250 mL HA solution of 15 mg/L was added in to the filtration cup, then the UV lamp was turned on. Samples were collected from the filtration cup using a syringe at intervals of 15 min for analysis. A blank sample using 0.05 g P25 (Degussa) was suspended in water and then deposited on the surface of a glass filter (0.45 μm). Based on our knowledge and experience, photo catalytic oxidation is guided by Langmuir-Hinshelwood (L-H) kinetics model equation-(1) as shown below. When the chemical concentration C_o is a millimolar solution (C_o small) the equation can be simplified to an apparent first-order equation-(2)[46]. A plot of $\ln C_o / C_t$ versus time will represent a straight line, the slope of which upon linear regression equals the apparent first-order rate constant k_{app}.

$$r = \frac{dC}{dt} = \frac{kKC}{1 + KC} \qquad (1)$$

$$Ln\left(\frac{C_o}{C_t}\right) = kKt = k_{app}t \qquad (2)$$

Where r is the oxidation rate of the reactant (mg/L.min), C the concentration of the reactant (mg/L), t the illumination time, k the reaction rate constant (mg/L.min), and K is the adsorption coefficient of the reactant (L/min).

The concentration of HA solution at 15 mg/L was filtered using the TiO_2 nanowire membrane and blank TiO_2 sample with/without UV irradiation. Its concentration and TOC in feed and filtrate were measured to calculate their respectively removal rates. The flux of TiO_2 nanowire membrane and blank TiO_2 sample was chosen at 4, 8, 12 and 16 L/min·m².

Results and discussion

Characterization of the membrane

The prepared TiO_2 nanowire membrane as shown in Figures 1 and 3 was a 47 mm-diameter membrane. The membrane fabrication had been described in the section of synthesis of TiO_2 nanowire membrane. This fabrication method has the following advantages:

(1) Homogeneity of the TiO_2 nanowire distribution can be formed through process of filtration.

(2) The membrane thickness can easily be controlled by adjusting the nanowire suspension concentration and volume of the suspension filtered.

(3) The TiO$_2$ nanowire membrane is highly flexible.

(4) Calcination above 300 °C ensures the membrane retained its physical shape. Its flexible property enables the membrane to be formed into various membrane modules for larger commercial applications.

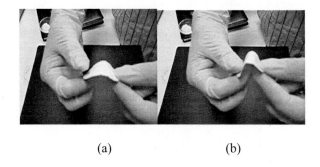

(a)   (b)

Figure 3. TiO$_2$ nanowire membrane flexibility.

FESEM images of TiO$_2$ nanowire membrane are shown in Figure 4. It can be seen that there is no significant effects on the structure of the TiO$_2$ nanowires after the calcination process from 300 to 700 °C. It also can be seen that TiO$_2$ nanowire membrane was formed by overlapping and interpenetrating of long nanowires with typical lengths in the range of several micrometers to several tens of a micrometer. The "spider web" nonwoven structure can be observed from Figure 4b. TEM image of TiO$_2$ nanowires membrane as shown in Figure 4c revealed that the length of these TiO$_2$ nanowires ranges from 20-100 nm in diameter.

(a)   (b)   (c)

Figure 4. TiO$_2$ nanowire membrane, (a) low magnification FESEM image, (b) high magnification FESEM image, (c) TEM image of the TiO$_2$ nanowire. (Figure 4a, 4b reproduced from reference, Zhang et al, Journal of Membrane Science 2008, 313, 44-51)

226

In order to determine the crystal structure of the nanowire membrane after calcination at different temperatures, the XRD was employed and the results are shown in Figure 5. Clearly, the fabricated nanowires using the method described earlier in section 2.1 are a mixture of anatase TiO_2 and titanate. The calcination temperature ranged from 300-500 °C, is able to convert TiO_2 and titanate into TiO_2-B phase and $Na_2(Ti_{12}O_{25})$, respectively. At the calcinations temperature of 700 °C, the TiO_2-B phase was then transformed into anatase and $Na_2(Ti_{12}O_{25})$. When temperature is increased further it will tend to decompose the $Na_2(Ti_6O_{13})$ and anatase TiO_2. The finding of these XRD results corresponds accordingly with the results reported [35, 42]. As Anatase TiO_2 has highest photocatalytic oxidation activity than the other phases of TiO_2, therefore TiO_2 nanowire membranes were calcinated at 700 °C.

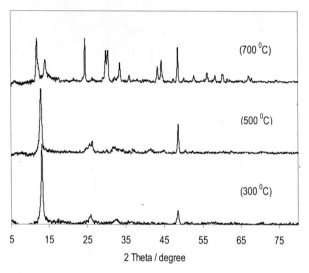

*Figure 5. XRD pattern of the TiO_2 nanowire membrane. (Reproduced from reference, Zhang et al, Journal of Membrane Science **2008**, 313, 44-51).*

To indentify the pore size distribution of the TiO_2 nanowire membrane, aqueous suspensions of polystyrene microspheres (0.05, 0.1, 0.2, 0.5 and 1μm in diameter, respectively), in the concentration of 0.033(wt)%, had been used. The separation factors of the TiO_2 nanowire membrane for these microspheres are shown in Figure 6. These results indicate that the separation factors for 0.2, 0.5, 1 and 2 μm polystyrene microspheres were more than 99%, indicating that these microspheres were unable to pass through the TiO_2 nanowire membrane. With the decrease of polystyrene microsphere diameter the separation factor for microspheres decreased as well. The separation factors for 0.1 and 0.05 μm microspheres were 96.3% and 89.5%, respectively, indicating that the TiO_2 membrane is capable of achieving partial removal of 0.1 and 0.05 μm microspheres. The pore size of a TiO_2 nanowire membrane can be categorized as the diameter of latex microspheres which are 90% retained by the membrane. Therefore, the pore size of the TiO_2 nanowire membrane is approximately about 0.05 μm.

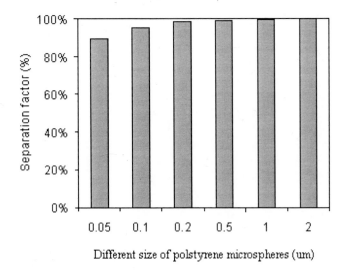

Figure 6. Separation factors of the TiO₂ nanowire membrane for different diameter polystyrene microspheres.

Photocatalytic oxidation activity for anti membrane fouling

In order to evaluate the multifunctional properties of the TiO_2 nanowire membrane, the photocatalytic activity of the TiO_2 nanowire membrane was studied in batch operation mode as described in section 2.3. HA under UV irradiation without P25 TiO_2 and TiO_2 nanowire membrane was also carried out as direct photolysis. P25 TiO_2 that was deposited on the glass filter was used as a reference. The removal rate of HA and TOC over the course of the three processes are shown in Figure 7. It can be seen clearly that the TiO_2 nanowire membrane shows excellent photocatalytic oxidation activity comparing with P25 TiO_2. The photocatalytic oxidation degradation of HA follows the first order kinetics which reveals that the apparent rate constant ($k_{app.}$) for the TiO_2 nanowire membrane was 0.022 min⁻¹ (R^2 = 0.97), a little lower than that of commercial P25 ($k_{app.}$ = 0.023 min⁻¹, R^2 = 0.98). The TOC removal rate curves as shown in Figure 7 also indicates that both TiO_2 nanowire membrane and P25 TiO_2 had a similar trend with a close TOC removal efficiency, hence the TOC removal in the solution indicates the mineralization of most HA into carbon dioxide and water. Compared with the photocatalytic oxidation degradation in the presence of either the TiO_2 nanowire membrane or P25 TiO_2, the degradation of HA by direct photolysis was very slow as the apparent rate constant was only 0.006 min⁻¹ which is one fourth of that of the photocatalytic degradations.

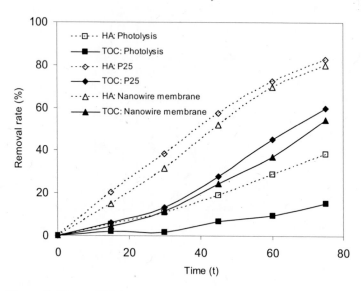

Figure 7. HA and TOC removal rates over the courses of photolysis, photocatalytic oxidation of HA via P25 deposited glass filter and TiO₂ nanowire membrane.

During TiO$_2$ nanowire membrane filtration, HA was deposited on the membrane surface and absorbed by the pore, thus forming a cake layer on the membrane surface and clogging of the membrane pore, which is called membrane fouling. Membrane fouling will result in increase of operation pressure or membrane flux decline. It is the main obstacle in membrane filtration as it causes the reduction in productivity and increase of operation cost. In contrast, the fouling in TiO$_2$ nanowire membrane does not exist. Under UV irradiation, the TiO$_2$ nanowires can be excited to generate high oxidative species, holes (h$^+$) and hydroxyl radicals (OH·). The mechanism of the photocatalytic oxidation degradation reactions is shown as follows:

$$TiO_2 + hv \ (\lambda < 387.4nm) \rightarrow TiO_2 \ (e^-_{CB} + h^+_{VB}) \tag{1}$$

$$h^+_{VB} + OH^- \rightarrow HO_{ads}\bullet \tag{2}$$

$$e^-_{CB} + O_2 \rightarrow O_2\bullet^- \tag{3}$$

$$O_2\bullet^- + H^+ \rightarrow HO_2\bullet \tag{4}$$

$$HO_2\bullet + O_2^-\bullet + H^+ \rightarrow O_2 + H_2O_2 \tag{5}$$

$$H_2O_2 + e^-_{CB} \rightarrow OH^- + HO\bullet \tag{6}$$

$$HO_{ads}\bullet + HA \rightarrow \text{mineralized products} \tag{7}$$

$$h^+_{VB} + HA \rightarrow \text{mineralized products} \tag{8}$$

It is clear that the main mechanism of photocatalytic oxidation reactions is the photogeneration of electron-hole pairs. Furthermore the photo generated holes (h^+) and electrons (e^-) migrate to the irradiated TiO_2 surface to act as adsorption sites or receptors for the organic compound and molecular oxygen, and take part in photo redox reactions in an air bubble reactor. More adsorption sites (receptors) may enrich the molecular oxygen and the humic acid on the TiO_2 surface. Hydroxyl (HO·) radicals formed from the protonation of the superoxide (O_2^-) anion (arising from adsorption of molecular oxygen on electron-rich sites) or the holes reaction with the electron donors in the electrolyte are highly reactive and powerful oxidant. These hydroxyl radicals may readily attack the chemisorbed HA ion to release aqueous organic anions, a surface carbonxylate radical and CO_2. In addition, the "hole" has oxidative properties, and is able to oxidize HA into CO_2.

The photocatalytic oxidation reactions for HA occur on the TiO_2 nanowires membrane surface so that the surface adsorption plays a critical role in photocatalysis and the rate of surface catalysis reaction needs to relate to the surface adsorption of the substrates.

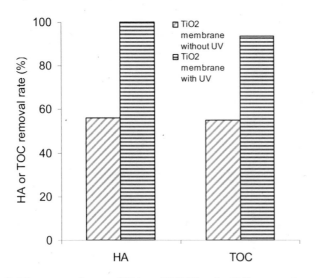

Figure 8. The removal rate of HA and TOC by the TiO_2 nanowire membrane filtration

As discussed above, HA can be effectively photocatalytical oxidation degraded into carbon dioxide and water or break down into small molecule weight matters, which cause no or less membrane fouling. Therefore, the fouling problem of TiO_2 nanowire membrane would be alleviated if a UV light was used to irradiate the TiO_2 nanowire membrane during the filtration. To investigate the anti-fouling ability of the TiO_2 nanowire membrane, HA solution of 15 mg/L was filtered using the TiO_2 nanowire membrane in a concurrent continuous operation mode in the experiment setup. The filtration without concurrent UV irradiation was also carried out as a reference while the flux of the membrane was kept at a constant of 4 L/min·m^2. The removal rates of HA

230

and TOC for the processes with/without UV irradiation are shown in Figure 8. Clearly, 57% of HA was rejected by the TiO$_2$ nanowire membrane alone without UV irradiation. While concurrent UV irradiation on TiO$_2$ nanowire membrane and filtration, the HA removal rate nearly reaches 100%. It also reveals that 93.6 % of TOC were removed by the TiO$_2$ nanowire membrane under the concurrent UV irradiation and filtration. These results shows that the TiO$_2$ nanowire membrane has the capability of self generation (anti-fouling).

Figure 9 shows the results of trans-membrane pressure (TMP) changes during filtration with/ without UV irradiation were monitored by a pressure sensor. It can be seen that the TMP of the TiO$_2$ nanowire membrane gradually increased with filtration time and increased rapidly after 7 h without UV irradiation. However, TMP of the TiO$_2$ nanowire membrane under concurrent UV irradiation and filtration was completely different from without UV irradiation. The TMP increased slightly at the initial stage and then kept at a constant during the rest of the filtration time. A constant TMP throughout the 30 h of filtration time revealed that the fouling of TiO$_2$ nanowire membrane did not occur. The TMP increased slightly at initial stage of filtration time probably due to the resistance of the TiO$_2$ nanowire membrane alone, which is identical to that of a conventional membrane.

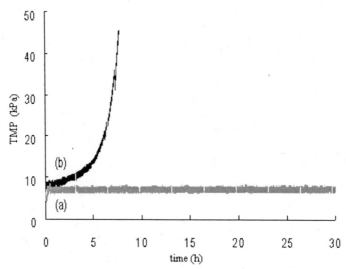

Figure 9. TMP changes of the TiO$_2$ nanowire membrane during the filtration. (a) with UV irradiation, (b) without UV irradiation. (Reproduced from reference, Zhang et al, Journal of Membrane Science 2008, 313, 44-51).

After photocatalytic oxidation and filtration, the TiO$_2$ nanowire membranes were further characterized using SEM. Figure 10 shows the images of the TiO$_2$ nanowire membrane surface and its cross section. Figure 10a clearly shows that in the absence of UV irradiation, a thick HA layer was deposited on the membrane surface, while a clean membrane surface was achieved (Figure 10b) in the presence of concurrent UV irradiation during filtration. There was not any HA cake layer found both on the membrane surface (Figure 10b) and inside of membrane pore (Figure 10c). These evidence (Figures 10a to 10c) clearly indicates that concurrent photocatalytic oxidation degradation has occurred on the TiO_2 nanowire membrane during its filtration and is able to effectively eliminated the HA fouling problem.

Fouling problems are the major obstacles to the conventional membrane used in the water industries. The consequence of fouling is reducing the membrane lifespan, generating new wastewater through membrane chemical cleaning thus increasing the cost of water production. These results presented in this section brings in excellent opportunities for the membrane water industries as membrane water purification can now be achieved in a cost effective manner if TiO_2 membrane can be implanted in the new future.

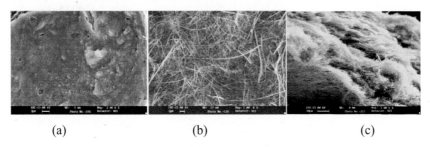

(a)   (b)   (c)

*Figure 10. The SEM images of TiO_2 nanowire membrane surface and cross section after filtration with and without UV irradiation. (a) the membrane surface without UV irradiation, (b) the membrane surface with UV irradiation, (c) the cross section with UV irradiation. Scale bars are 3 μm for (a) and (b) and is 10 μm for (c).(Reproduced from reference, Zhang et al, Journal of Membrane Science **2008**, 313, 44-51).*

Figures 7 to 10 clearly reveal that the free-standing, flexible and multifunctional TiO_2 nanowire membrane is able to perform the concurrent filtration and photocatalytic oxidation for not only removing the HA from water, but also able to oxidize the HA foulant into CO_2, thus membrane self-regenerates (anti-fouling).

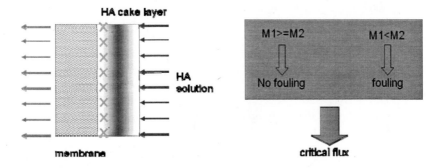

Figure 11. Mechanism of concurrent TiO_2 nanowire membrane filtration and its photocatalytic oxidation anti-fouling

The photocatalytic oxidation capacity of the TiO_2 nanowire membrane, the maximal amount of HA photocatalytically oxidized per unit time (M1), is a function of membrane area and light intensity on the membrane surface. During filtration, the amount of HA rejected on membrane surface per unit time (M2) is a function of flux and HA concentration. When M2 is less than M1, the HA rejected on membrane surface could be completely photocatalytical oxidation degraded so that the TiO_2 nanowire membrane fouling would be eliminated. Whereas, when M2 is greater than M1, the HA rejected on TiO_2 nanowire membrane surface could not be completely degraded. Therefore, the HA would deposit and absorb on and within the TiO_2 nanowire membrane surface and pores respectively, resulting in the membrane fouling. When the concentration of HA in feed is constant, M2 is a function of only membrane flux. Fouling would not occur when operating below the critical flux of TiO_2 nanowire membrane. Figure 11 illustrate the mechanism of concurrent TiO_2 nanowire membrane filtration and its photocatalytic oxidation anti-fouling.

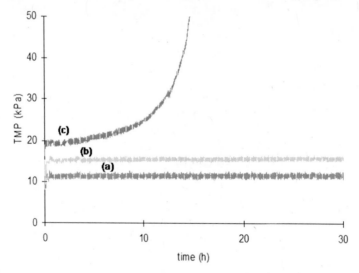

Figure 12. TMP changes of the TiO_2 nanowire membrane filtration at different flux. (a) 8 L/min.m², (b) 12 L/min.m², (c) 16 L/min.m²

In order to verify the mechanism of concurrent filtration and photocatalytic oxidation anti-fouling as illustrated in Figure 11, the continuous filtration of 15 mg/L HA using the TiO_2 nanowire membrane under UV irradiation was carried out at three different fluxes, 8, 12 and 16 L/min.m². The TMP was monitored during filtration time to determine the membrane fouling. The results are shown in Figure 12, clearly reveal that the TMP gradually increases with the filtration time and rapidly increase with filtration time after 12 h of filtration at 16 L/min.m², which indicates the increasing HA deposition and adsorption on membrane surface and within pore, resulting in membrane fouling. It also can be observed that the TiO_2 nanowire membrane fouling behavior is similar to that of the conventional filtration membrane. In contrast, constant TMP at flux of 8 and

12 L/min.m^2 were observed, indicating no membrane fouling had occured. Therefore 12 L/min.m^2 is the critical flux of the concurrent filtration and TiO$_2$ nanowire membrane at a concentration of 15mg/L HA.

Conclusion and Perspectives

Conclusion

The free-standing, flexible and multifunctional TiO$_2$ nanowire membrane in the form of a "spider web" nonwoven structure was fabricated using a simple hydrothermal-filtration method. The TiO$_2$ nanowire membrane showed an excellent performance on concurrent filtration and photocatalytic oxidation degradation of HA in water. Nearly, 100% HA and 93.6% TOC removal were achieved in the concurrent continuous filtration and photocatalytic oxidation processes. 93.6% TOC removal indicates that HA has been completely mineralized into CO$_2$ and water instead of intermediates matter. The experimental results revealed that TiO$_2$ nanowire membrane fouling caused by HA deposition and adsorption was alleviated by concurrent photocatalytic oxidation degradation. We believe that the free-standing, flexible and multifunctionalTiO$_2$ nanowire membrane will have a good potential for wide application in water and wastewater treatment and can be extended to eliminate other foulants such as microorganisms and trace organics.

Perspectives

New and increasingly stringent regulations for drinking water require a more reliable and sustainable membrane treatment technology. The free-standing, flexible and multifunctional TiO$_2$ nanowire membrane has the potential to be applied for large scale water and wastewater treatment, as well as desalination. It would be able to overcome the existing problems of conventional membrane fouling resulting in a lower water production cost and higher water quality.

The ability to remove the living microorganisms from Ballast water will be radically transformed by development of novel materials. The emerging ability of using the free-standing, flexible and multifunctional TiO$_2$ nanowire membrane to remove living microorganisms from Ballast water can be expected to lead an entirely new type of cost effective on-board treatment technology which will be able to overcome the existing problems of Ballast water discharge for the marine industry. The module operation with its foot-print size makes it feasible to commercial application.

Since the phenomenon of photoelectrochemical hydrogen generation was first discovered by Fujishima and Honda reported in Nature dated 1972 [47], TiO$_2$ research has attracted many researchers' attention, large amount of research efforts have been carried out particularly in using TiO_2. Using the free-standing, flexible and multifunctional TiO_2 nanowire membrane, we should be able to design a new membrane reactor system which would be comparably easier for commercial operation to that of a TiO_2 particle reactor system and it too could provide a new source of hydrogen; an interesting possibility for clean energy industries.

Until today, solar cell (photovoltaics), the conversion of sunlight to electricity has been dominated by solid-state junction devices, often made of silicon. The highest energy conversion efficiency obtained from lab studies is 28%, makes it feasible to be widely used in commercial application. However, the high cost of its fabrication of the silicon based solar cell and non- flexible property prohibits its wide application in the large commercial field. It is expected that using the free-standing, flexible and multifunctional TiO_2 nanowire membrane could produce a new flexible dye sensitized solar cell (DSSC) which could generate electricity at a very low cost; renewable energy from a systematically controlled solar cell.

As the results shown in this section, it is unprecedented to have such nanostructured TiO_2 nanowire. When the TiO_2 nanowire is assembled into a free-standing, flexible and ultrathin miltifunctional membranes, it could provide a solution for a variety of applications, including drug delivery, self cleaning glass and bathtub, and a "spider web" non-woven textiles which are stable at high-temperatures and chemical inertness etc. We believe that other kind of nanostructured TiO_2 materials such as TiO_2 nanotube and nanofiber should be able to be fabricated and used for assembling the membrane for not only drinking water production and but also for energy production as well.

Reference

1. Kolpin, D.W.; Furlong, E.T.; Meyer, M.T.; Thurman E.M.; Zaugg, S.D.; Barber, L.B.; Buxton, H.T. Pharmaceuticals, Hormones, and Other Organic Wastewater Contaminants in U.S. Streams, 1999-2000: A National Reconnaissance, *Environ. Sci. Technol.* **2002**, 36 1202-1211.

2. World Health Organization, in "Addendum to Vol.1, Recommendations, WHO, Geneva, 1998.

3. USEPA Drinking Water Priority Rulemaking: Microbial and Disinfection By product Rules", in Drinking Water Standard, 1998

4. Cho, J.; Amy, G.; Pellegrino, J. Membrane filtration of natural organic matter: factors and mechanisms affecting rejection and flux decline with charged ultrafiltration (UF) membrane, *Journal of Membrane Science* **2000**, 164, 89-110.

5. Yuan, W.; Zydney, A.L. Humic acid fouling during microfiltration, *Journal of Membrane Science* **1999,** 157, 1-12.

6. Richardson, S.D.; Thruston, A.D.; Rav-Acha, C.; Groisman, L.; Popilevsky, I.; Juraev, O.; Glezer, V.; McKague, A.B.; Plewa, M.J.;

Wagner, E.D. Tribromopyrrole, Brominated Acids, and Other Disinfection Byproducts Produced by Disinfection of Drinking Water Rich in Bromide, *Environ. Sci. Technol.*, **2003**, 37, 3782-3793.

7. Huang, W.-J.; Chen, L.-Y.; Peng, H.-S. Effect of NOM characteristics on brominated organics formation by ozonation, *Environment International* **2004**, 29, 1049-1055.

8. Sohn, J.; Amy, G.; Cho, J.; Lee, Y.; Yoon, Y. Disinfectant decay and disinfection by-products formation model development: chlorination and ozonation by-products, *Water Research* **2004** 38,2461-2478.

9. Krasner, S.W.; Weinberg, H.S.; Richardson, S.D.; Pastor, S.J.; Chinn, R.; Sclimenti, M.J.; Onstad, G.D.; Thruston, A.D. Occurrence of a New Generation of Disinfection Byproducts, *Environ. Sci. Technol.* **2006** 40, 7175-7185.

10. Hoffmann, M.R.; Martin, S.T.; Choi, W.; Bahnemann, D.W. Environmental Applications of Semiconductor Photocatalysis, *Chem. Rev.*, **1995**, 95, 69-96.

11. Fujishima, A.; Rao, T.N.; Tryk, D.A. Titanium dioxide photocatalysis, *Journal of Photochemistry and Photobiology C: Photochemistry Reviews* **2000**, 11-21.

12. Herrmann, J.-M. Heterogeneous photocatalysis: fundamentals and applications to the removal of various types of aqueous pollutants, *Catalysis Today* **1999**, 53, 115-129.

13. Li, X.Z.; Liu, H.; Cheng, L.F.; Tong, H.J. Photocatalytic Oxidation Using a New Catalyst-TiO_2 Microsphere-for Water and Wastewater Treatment, *Environ. Sci. Technol.* **2003** 37,3989-3994.

14. Zhang, X.; Wang, Y.; Li, G. Effect of operating parameters on microwave assisted photocatalytic degradation of azo dye X-3B with grain TiO_2 catalyst, *Journal of Molecular Catalysis A: Chemical*, **2005** 237,199-205.

15. Anderson, M.A.; Gieselmann, M.J.; Xu, Q. Titania and alumina ceramic membranes, *Journal of Membrane Science* **1988**, 39, 243-258.

16. Moosemiller, M.D.; Hill, C.G.; Anderson, M.A. Physicochemical Properties of Supported γ-Al_2O_3 and TiO_2 Ceramic Membranes, *Separation Science and Technology* **1989**, 24 641 - 657.

17. Choi, H.; Stathatos, E.; Dionysiou, D.D. Photocatalytic TiO_2 films and membranes for the development of efficient wastewater treatment and reuse systems, *Desalination*, **2007**, 202, 199-206.

18. Triani, G.; Evans, P.J.; Attard, D.J.; Prince, K.E.; Bartlett, J.; Tan, S.; Burford, R.P. Nanostructured TiO_2 membranes by atomic layer deposition, *Journal of Materials Chemistry* **2006**, 16,1355-1359.

19. Meulenberg, W.A.; Mertens, J.; Bram, M.; Buchkremer, H.-P.; Stover, D. Graded porous TiO_2 membranes for microfiltration, *Journal of the European Ceramic Society* **2006**, 26, 449-454.

20. Ding, X.; Fan, Y.; Xu, N. A new route for the fabrication of TiO_2 ultrafiltration membranes with suspension derived from a wet chemical synthesis, *Journal of Membrane Science* **2006**, 270,179-186.

21. Choi, H.; Sofranko, A.C.; Dionysious, D.D. Nanocrystalline TiO_2 Phototatalytic Membranes with a Hierarchical Mesoporous Multilayer

236

Structure: Synthesis, Charaterization, and Multifunction, *Advanced Functional Materials* **2006**, 16, 1067-1074.

22. Bosc, F.; Lacroix-Desmazes, P.; Ayral, A. TiO$_2$ anatase-based membranes with hierarchical porosity and photocatalytic properties, *Journal of Colloid and Interface Science* **2006,** 30, 4545-548.

23. Nishino, J.; Information, C.; Teekateerawej, S.; Nosaka, Y. Preparation of TiO$_2$ coated Al$_2$O$_3$ membrane by a pyrolysis method, *Journal of Materials Science Letters* **2003**, 22, 1007-1009.

24. Ha, H.Y.; Nam, S.W.; Lim, T.H.; Oh, I.-H.; Hong, S.-A. Properties of the TiO$_2$ membranes prepared by CVD of titanium tetraisopropoxide, *Journal of Membrane Science* **1996**, 11, 181-92.

25. Van Gestel, T.; Vandecasteele, C.; A. Buekenhoudt, C. Dotremont, Luyten, J.; Leysen, R.; Van der Bruggen, B.; Maes, G. Alumina and titania multilayer membranes for nanofiltration: preparation, characterization and chemical stability, *Journal of Membrane Science* **2002**, 207, 73-89.

26. Yang, W.P.; Huang, S.-L. Synthesis and Characterization of Titania Membrane, *Separation Science and Technology* **2003**, 38, 4027-4040.

27. Zeng, Z.-q.; Xiao, X.-y.; Gui, Z.-l.; Li, L.-T. Al$_2$O$_3$-SiO$_2$-TiO$_2$ composite ceramic membranes from sol-gel processing, *Materials Letters*, **1998**, 3567-71.

28. Molinari, R.; Palmisano, L.; Drioli, E.; Schiavello, M. Studies on various reactor configurations for coupling photocatalysis and membrane processes in water purification, *Journal of Membrane Science* **2002**, 206, 399-415.

29. Molinari, R.; Mungari, M.; Drioli, E.; Di Paola, A.; Loddo, V.; Palmisano, L.; Schiavello, M. Study on a photocatalytic membrane reactor for water purification, *Catalysis Today* **2000** 55, 71-78.

30. Jung, J.H.; Kobayashi H.; van Bommel, K.J.C.; Shinkai, S.; Shimizu, T. Creation of Novel Helical Ribbon and Double-Layered Nanotube TiO$_2$ Structures Using an Organogel Template, *Chem. Mater.*, **2002** 14, 1445-1447.

31. Yao, B.D.; Chan, Y.F.; Zhang, X.Y.; Zhang, W.F.; Yang, Z.Y.; Wang, N. Formation mechanism of TiO$_2$ nanotubes, *Applied Physics Letters* **2003**, 82, 281-283.

32. Kasuga, T.; Hiramatsu, M.; Hoson, A.; Sekino, T.; Niihara, K. Formation of titanium oxide nanotube, *Langmuir*, **1998** 14, 3160-3163.

33. Tian, Z.R.; Voigt, J.A.; Liu, J.; McKenzie, B.; Xu, H. Large Oriented arrays and continuous films of TiO$_2$-based nanotubes, *J. Am. Chem. Soc.* **2003**, 125 12384-12385.

34. Yoshida, R.; Suzuki, Y.; Yoshikawa, S. Syntheses of TiO$_2$(B) nanowires and TiO$_2$ anatase nanowires by hydrothermal and post-heat treatments, *Journal of Solid State Chemistry*, **2005**, 178, 2179-2185.

35. Yuan, Z.-Y.; Su, B.-L. Titanium oxide nanotubes, nanofibers and nanowires, *Colloids and Surfaces A: Physicochemical and Engineering Aspects*, **2004**, 241, 173-183.

36. Chen, Y.; Crittenden, J.C.; Hackney, S.; Sutter, L.; Hand, D.W. Preparation of a novel TiO_2-based p-n junction nanotube photocatalyst, *Environ. Sci. Technol.* **2005**, 39, 1201-1208.

37. Chen, X.; Mao, S.S. Titanium dioxide nanomaterials: Synthesis, properties, modifications, and applications, *Chem. Rev.* **2007**, 1072891-2959.

38. Gu, G.; Schmid, M.; Chiu, P.-W.; Minett, A.; Fraysse J.; Kim, S.R. G.-T; Kozlov, M.; Munoz, E.; Baughman. R.H. V_2O_5 nanofibre sheet actuators, *Nature Materials* **2003**, 2, 316-319.

39. Wu, Z.; Chen, Z.; Du, X.; Logan, J.M.; Sippel, J.; Nikolou, M.; Kamaras, K.; Reynolds, J.R.; Tanner, D.B.; Hebard, A.F.; Rinzler, A.G. Transparent, conductive carbon nanotube films, *Science* **2004**, 305, 1273-1276.

40. Endo, M.; Muramatsu, H.; Hayashi, T.; Kim, Y.A.; Terrones, M.; Dresselhaus M.S. Nanotechnology: "Buckypaper" from coaxial nanotubes, *Nature* **2005** 433, 476-476.

41. Zhang, M.; Fang, S.; Zakhidov, A.A.; Lee, S.B.; Aliev, A.E.; Williams, C.D.; Atkinson, K.R.; Baughman, R.H. Strong, transparent, multifunctional, carbon nanotube sheets, *Science* **2005**, 309 1215-1219.

42. Dong, W.; Cogbill, A.; Zhang, T.; Ghosh, S.; Tian, Z.R. Multifunctional, catalytic nanowire membranes and the membrane-based 3D devices, *J. Phys. Chem. B* **2006**, 110, 16819-16822.

43. Armstrong, A.R.; Canales, G.A.J.; Bruce, P.G. TiO_2-B nanowires, angewandte chemie international edition, 2004, 43, 2286-2288.

44. Nakao, S.-i. Determination of pore size and pore size distribution : 3. Filtration membranes, *J. Mem. Sci.* **1994**, 96131-165.

45. Singh, S.; Khulbe, K.C.; Matsuura, T.; Ramamurthy, P.; Membrane characterization by solute transport and atomic force microscopy, *J. Mem. Sci.* **1998**, 142, 111-127.

46. Konstantinou, I. K.; Albanis, T.A. TiO_2-assisted photocatalytic degradation of azo dyes in aqueous solution: kinetic and mechanistic investigations: A review, *App. Catal. B*, **2004**, 49, 1-14.

47. Fujishima, A. and Honda K., Electrochemical photolysis of water at a semiconductor electrode. *Nature,* **1972**, 238, 37.

Indexes

Author Index

Subject Index